# ANIMAL FEEDS

# ANIMAL FEEDS

Nekesah T. Wafullah

Delve
PUBLISHING

www.delvepublishing.com

**Animal Feeds**

*Nekesah T. Wafullah*

**Delve Publishing**
4164 Lakeshore Road
Burlington, ON L7L 1A4
Canada
www.delvepublishing.com

Email: orders@arclereducation.com



ISBN: 978-1-77956-401-6 (Hardcover)

# ABOUT THE AUTHOR

**Nekesah T. Wafullah** is a skilled agriculture expert with extensive knowledge in agricultural energy value addition products, agricultural business management services, project management, various forms of fertilizer, their production, sales, marketing aspects and application regimes; cross border fertilizer trade policies; youth and women empowerment and volunteerism. She is adept at project planning and management as well as creating simple solutions to complex problems. She has experience within agricultural markets in Kenya, Zambia, Malawi, Tanzania, Rwanda, the Democratic Republic of Congo (DRC)- Bukavu and Lubumbashi and Uganda. She mentors high school and college students and advocates for better performance in Agricultural science. During her free time, she loves editing books, watching movies, cooking, baking, networking, reading and dancing. Nekesah holds an M Sc. In Agricultural and Applied Economics degree from the University of Nairobi with a major in International Trade and Policy.

# TABLE OF CONTENTS

# LIST OF FIGURES

# LIST OF TABLES

# LIST OF ABBREVIATIONS

| | |
|---|---|
| ATP | Adenosine Triphosphate |
| BCS | Body Condition Scoring |
| BMR | Basal Metabolic Rate |
| BSG | Brewer's Spent Grains |
| BUN | Blood Urea Nitrogen |
| CAGR | Compound Annual Growth Rate |
| CCPs | Critical Control Points |
| CLA | Conjugated Linoleic Acid |
| CoA | Acetyl Coenzyme A |
| CP | Crude Protein |
| CSB | Condensed Separator Byproduct |
| DDGS | Distiller's Dried Grains with Soluble |
| DE | Digestion-Efficient Energy |
| DIP | Degradable Intake Protein |
| DM | Dry Matter |
| DNA | Deoxyribonucleic Acid |
| EU | European Union |
| FAD | Flavin Adenine Dinucleotide |
| FAO | Food and Agriculture Organization |
| FEFAC | The European Feed Manufacturers' Federation |
| GAP | Good Agricultural Practices |
| GE | Gross Energy |
| GI | Gastrointestinal Tract |
| GMP | Good Manufacturing Practices |
| HACCP | Hazard Analysis and Critical Control Point |
| IFFO | International Fishmeal and Fish Oil Organization |
| LP | Linear Programming |
| ME | Metabolizable Energy |
| MRL | Maximum Residue Limits |

| | |
|---|---|
| NAD | Nicotinamide Adenine Dinucleotide |
| NE | Net Energy |
| NPN | Non-Protein Nitrogen |
| NPU | Net Protein Utilization |
| NSP | Non-starch Polysaccharides |
| PAP | Processed Animal Protein |
| PTH | Parathyroid Hormones |
| PUFAs | Polyunsaturated Fatty Acids |
| RNA | Ribonucleic Acid |
| RUP | Rumen Undegradable Protein |
| SCP | Single Cell Proteins |
| SOPs | Standard Operating Procedure |
| TCA | Tricarboxylic Acid |
| TDN | Total Digestible Nutrients |
| TEF | Toxicity Equivalency Factor |
| UIP | Undegraded Intake Protein |
| VFAs | Volatile Fatty Acids |
| WHO | World Health Organization |

# ACKNOWLEDGMENT

The book is the product of great effort and time. Its completion would not have been possible without the participation and assistance of many people, whose names cannot all be enumerated. Their contributions are sincerely appreciated and gratefully acknowledged. However, I would like to express my deep appreciation and indebtedness particularly to the following: Arcler Education, Inc. and James M. Kihungi, for their endless support, kindness, and understanding spirit during this undertaking. To all relatives, friends, and others who, in one way or another, shared their support—whether morally, financially, or physically—I would like to personally thank you.

# PREFACE

The animal feed sector plays an important role in maintaining the best possible nutrition and overall health for livestock. For the best output of animal products, different sources of feed are used for the nutritional management of livestock. In general, animals need the same nutrients that people do. Certain feeds are grown expressly for animals, including cereal grains, hay and silage crops, and pasture grasses. After a food crop is processed for human use, by-products such as sugar beetroot pulp, brewers' grains and pineapple fiber are left over. Basic nutrient needs for animals include carbs, protein, fat, minerals, vitamins, and water. Protein is also necessary for the development of the muscles and other body parts in juvenile animals. Carbohydrates and lipids, provide energy to most animals. Minerals and vitamins serve different bodily functions. Despite the need for the mentioned nutrients, side effects are experienced in case of deficiencies or excesses. While deficiencies cause disorders/diseases, excesses on the other hand lead to toxicity. To ensure that animals are obtaining the required nutritive amount, valuation of nutrients is important. One thing that stands out, is that the quality of nutrients varies with different sources. Therefore, it is important to consider sourcing feed from high-quality origins.

Animal feeds are classified into roughages and concentrates. Roughages consist of root crops, straw, hays, silage, pasture grasses, and stover (cornstalks), while concentrates include high-energy-value concentrates, such as fat, high-protein oil meals or cakes, and by-products from processing. Apart from traditional feed sources, newer alternatives have emerged, such as insect meals, microalgae, and seaweed or macroalgae. Based on livestock adaptation, feeds provided to the animal are utilized through processes such as the glycolysis pathway, the Krebs cycle, the urea cycle, and fatty acid biosynthesis.

Apart from feed that provides basic nutrients, feed additives are added in small rations to enhance the nutritional value and visual appeal of food, as well as the quality and digestibility of feed, as well as the health and performance of the animals. In animal rations, enzymes, pro- and prebiotics, antioxidants, antibiotic growth promoters, and coloring compounds are some of the feed additives that are most frequently utilized. The combined goals of these various ingredients are to improve animal gut health (pro/prebiotics, for example) and food product quality (antioxidants), decrease nutrient loss (phosphorus, nitrogen, etc.), increase the availability or digestibility of bound nutrients (enzymes), and support environmental protection.

To achieve the required animal nutritional needs sources such as forage, for example, alfalfa, straw, hay, and silage are used. Additionally, concentrates such as ammonium sulfate cereals and other industry co-products are used. To ensure, nutrient availability, reduction of anti-nutritional factors, preservation, and improved appearance and palatability processing is necessary for most of the feedstuff. The processing techniques include grinding, chopping, mixing, pelleting, and baling. Most of the processes require appropriate mill equipment for efficient feed production. During processing, it is important to ensure feed safety, through the application of the Hazard Analysis and Critical Control Point (HACCP) concepts at farm-level processing or large industry production. HACCP principles should be observed at all points of handling animal feeds to ensure quality and safety.

—***Author***

Chapter 1

# Introduction to Animal Feeds

## Contents

## 1.1. INTRODUCTION

To maintain the best possible nutrition and overall health for livestock and poultry, the animal feed sector plays a crucial role in global agriculture. Although it is more explicitly used to refer to animal fodder, the word "animal feed" can be used to refer to any food provided to domestic animals farmed for their products or as home pets. In order to create extremely nutritious diets that both preserve the health of the animals and improve the quality of such byproducts as meat, milk, or eggs, modern feeds are made by carefully choosing and combining ingredients. Agricultural scientists' study, testing, and chemical analysis have led to continuous advances in animal feeds. Unlike fodder, which is fed to animals in the field and pasture, fodder is food that is specifically grown for animals and is transported and delivered to them.

The majority of animal feed is produced for fish, poultry, pigs, and cattle. Animal feed is primarily made from field cereals, particularly soybeans and maize. Forage material is used to make hay, straw, and silage, which are all regarded as types of animal feed. Traditional sources of animal feed also include leftover food from brewing, milling, and ethanol production, as well as home garbage. Remaining residue after milling maize, soy, and peanuts can be a valuable source of animal feed. The main waste product from brewing, known as brewer's spent grains (BSG), is fed to animals. Distiller's Dried Grains with Soluble (DDGS) are leftovers from the production of ethanol used in both alcohol and petrol. In recent years, DDGS production has expanded significantly, and it is now a productive substitute for soybean and maize meal. Compared to corn and soybean meals, DDGS has a longer shelf life and offers energy that is similarly metabolized and digestible. Pork diets are the main application for it; many producers use 20% DDGS. Concentrates and roughages are two fairly broad definitions of animal feed kinds. Cereal grains, byproducts, and oil meals or cakes are examples of concentrates. They have a lot of energy. Grains such as corn, oats, barley, rye, sorghum, and wheat are used to make concentrates. Canola, peanut, cottonseed, and soybean oil are examples of oil meals or cakes. Byproducts from the processing of fish, livestock, sugarcane, sugar beets, and both BSG and DDGS can be utilized. Grass pasture or plant remnants like silage, straw, cornstalks, grasses, and hay are referred to as roughages. Concentrates are fed to fish, swine, and poultry. Roughages make up the majority of the food needed to meet the digestive needs of livestock ruminants.

In general, animals need the same nutrients that people do. Certain feeds are grown expressly for animals, including several cereal grains, hay and silage crops, and pasture grasses. After a food crop is processed for human use, by-products such as sugar beetroot pulp, brewers' grains and pineapple fiber are left over. These feeds are known as alternative feeds. Animals may also be fed excess food crops, such as roots, fruits, vegetables, wheat, and other cereals. The beginning of the use of dried roughage and other preserved meals for animals is not documented in history. The majority of historical accounts speak about individuals who were nomads and who followed the natural feed sources with their flocks and herds. It's likely that some of the leftovers were fed to domesticated animals while they were employed in agricultural production. Albrecht von Thaer, a German agriculturist, undoubtedly made the first attempt in science to compare feeds for animals in 1809 when he created "hay values," which are metrics for feeds' nutritional value. Following, tables detailing the value of feeds and animal requirements in Germany were adopted by other nations. In northern Europe, it has long been customary to store green forages like beetroot leaves and maize (maize) plants into earthen pits for preservation. To preserve and use more of the corn plant, the concept of silage production was progressively developed in Europe and brought to the United States in the 1870s from France. Much of the coarse stem was wasted when the mature, dried maize plant was fed to calves in the winter, but everything was used when it was chopped and ensiled (formed into silage). Concrete bunker silos used to store silage were ubiquitous in many rural locations across the world during the 20$^{th}$ century.

## 1.2. BASIC NUTRIENTS AND ADDITIVES

Carbs, protein, fat, minerals, vitamins, and water are the essential elements that animals need for upkeep, development, reproduction, and overall health. Carbohydrates and lipids provide the majority of the energy required for growth and exercise. Additionally, protein can provide energy, especially if the body isn't getting enough fat and carbohydrates or if the body is consuming more protein than it requires. Animals require a source of energy to maintain internal bodily functions and to power their muscles. When an animal consumes more energy than it needs, the extra energy is stored as body fat, which it can use as fuel in the event that food becomes scarcer.

### 1.2.1. Proteins

Protein is also necessary for the development of the muscles and other body parts in juvenile animals. Since milk, eggs, and wool are high in protein, the animals that produce them require extra protein in their diet. For maintenance, or the everyday mending of muscles, internal organs, and other bodily tissues, all animals need a small amount of protein. More than 20 distinct amino acids are found in proteins, and these acids are released during digestion. Humans, monkeys, pigs, poultry, rabbits, and mink are among the animals with a simple single stomach (monogastric), and they need the necessary daily amounts of the 10 essential amino acids: arginine, histidine, isoleucine, leucine, lysine, methionine, phenylalanine, threonine, tryptophan, and valine. Poultry also require glutamic acid and glycine for growth in addition to these. Up to half of the methionine needs and half of the phenylalanine requirement can be satisfied by substituting tyrosine with cysteine. High-quality protein has high concentrations of the necessary amino acids in the right ratio for optimal utilization. Examples of this type of protein include eggs, milk, fish meal, meat by-products, and soybean meal. Protein of low quality, like that found in most grains like corn, barley, and sorghum, has insufficient amounts of one or more vital amino acids. When combined with other feeds that restore the proper balance of critical amino acids, meals with lower-quality proteins can still be beneficial. For ruminants—cattle, sheep, goats, and other animals with four stomachs—the amino acid profile of a protein source is secondary because the bacteria that aid in food digestion in the rumen—the first stomach—use simple nitrogen compounds to build proteins in their cells. The animals break down the germs further along the digestive tract. Through this indirect process, urea (a nitrogen molecule) or a meal that may have initially contained low-quality protein is converted by ruminants into high-quality protein. Calves, lambs, and young ruminants, on the other hand, require high-quality protein until the rumen matures to the point where this bacterial process becomes established.

### 1.2.2. Carbohydrates and Fats

Carbohydrates and lipids, which are oxidized in the body, provide energy to most animals. These release heat, which keeps the body at a constant temperature, provides energy for muscular contraction and growth and keeps essential processes going. For development, work, or milk production, animals require a lot more energy (and feed overall) than they do for basic maintenance. All animals can digest simple carbohydrates like sugars and starches with ease. In the rumen of sheep and cattle, or in the cecum of rabbits and horses, bacteria, and protozoa break down the complex carbohydrates (cellulose, hemicelluloses) that comprise the fibrous stems of plants. Dogs, cats, birds, and lab animals cannot digest such complex carbohydrates to any significant degree, nor can humans. Therefore, compared to monogastric carnivores and omnivores, for which fibrous materials have little to no energy value, ruminants, and certain herbivorous animals receive a far greater amount of the nutrients that provide energy from the carbohydrates of plants. Because it is readily digested and provides roughly 2.5 times the energy of an equivalent weight of grain or sugar, fat in feeds has a high nutritional value. Except trace amounts of necessary fatty acids, fat can be substituted in the diet for an equivalent amount of digestible carbohydrates, despite its high nutritional value. For growth and health, very modest levels of linoleic acid, an unsaturated fatty acid found in certain fats, are required. This acid is usually present in sufficient proportions in animal feeds unless processing has eliminated it.

### 1.2.3. Minerals

Common salt (sodium chloride), calcium, phosphorus, sulfur, potassium, magnesium, manganese, iron, copper, cobalt, iodine, zinc, molybdenum, and selenium are among the minerals that are necessary for animal life. If an excessive amount is consumed by animals, the latter six of these can be poisonous. Generally speaking, all farm animals require more common salt than what is included in their meals, and they are routinely given it. Because they are most intensively used in the production of bones, milk, and eggshells, phosphorus, and calcium are the other important minerals that are most likely to be deficient. Bonemeal, dicalcium phosphate, and defluorinated phosphates are good sources of both calcium and phosphorus. Almost all of an eggshell is calcium carbonate. Ground limestone, ground seashells, and marl are all rich sources of calcium that can be easily obtained. Animals require trace amounts of iodine to make the thyroid gland's released thyroxine, an iodine-containing molecule. A severe iodine deficit can result in goiter, a condition where the thyroid gland enlarges dramatically. Goiter has resulted in significant losses of young piglets, lambs, children, calves, and foals in some areas. If the mother is given iodized salt before the offspring are born, iodine deficits can be avoided. Iodine is a common addition to almost all commercial sources of animal salt. Some soils and forages lack copper and cobalt, which are essential for the synthesis of hemoglobin, the pigment that carries oxygen in red blood cells, along with iron. Farm animals in these areas may experience anemia if the lack is not remedied with an appropriate mineral supplement. Typically, cobalt is given to ruminants through feed so they can subsequently synthesize vitamin B12. Pigs and other monogastric animals need to eat a direct source of vitamin B12. With the exception of milk, most animal feeds are rich in iron, which is necessary for the production of hemoglobin. When young suckling piglets don't start eating other feeds than milk, the only real issue with iron deficiency arises. To achieve their iron needs, they need to be injected with iron or have access to new soil. Animals require manganese, and all agricultural animals get enough of it from their regular meals. The absence of manganese can result in sliding tendon (perosis), a nutritional illness that affects chicks and young turkeys, as well as egg failure. Swine typically eat diets low in zinc, particularly when there is an abundance of calcium present. Zinc deficiencies can be prevented by fortifying feed with 100 parts per million of zinc, such as zinc sulphate or zinc carbonate. This will avoid symptoms such as stunted growth and severe skin scaliness and cracking (parakeratosis). While selenium is required in trace amounts for normal health, excessive amounts—which are poisonous and can even be fatal—can be found in some grains and forages. Livestock that is grazing may be given unrestricted access to a combination consisting of 60% dicalcium phosphate and 40% common salt to provide them with both calcium and phosphorus. When there may be a deficiency in copper or cobalt, trace mineralized salt is utilized. To prevent them from having to consume more other minerals than necessary to obtain the necessary amount of salt, livestock are typically allowed access to common salt on their own. Typical swine diets include the recommended amounts of salt, calcium, phosphorus, and other trace minerals, some of which may be lacking in the grains they are fed.

### 1.2.4. Vitamins

The water-soluble B group comprising thiamin, riboflavin, niacin, pantothenic acid, choline, biotin, folic acid, and vitamins B6 and B12 as well as vitamin C are known vitamins. The fat-soluble vitamins A, D, E, and K are other known vitamins. The vitamin most likely to be deficient in animal diets is vitamin A, which is necessary for development, reproduction, production of milk, and the preservation of a healthy defense against respiratory diseases. Carotene, which plants that grow in the green, is abundant and may be converted by animals into vitamin A. In situations where animals are not fed green pastures or given adequate pasture, vitamin A supplements are added to their meals to ensure their supply. Animals need vitamin D to utilize calcium and phosphorus; in young, developing animals, a vitamin D shortage results in rickets. Sunlight's UV rays convert the provitamin in the skin into vitamin D. Hay that has been field cured produces vitamin D because sunshine reacts with the ergosterol in the hay crops. Vitamin D content in some fish oils is extremely high. Vitamin D is abundant in livestock that spend a lot of time outside in the sun. Cattle, sheep, and horses often receive sufficient quantities from the hay they are fed during the winter months in cold climates; indoor-raised laboratory animals, pigs, and poultry will not receive enough unless a supplement is added. Since the microorganisms in their rumen produce these vitamins, the vitamin B group is not necessary for feeding sheep, cattle, or other ruminants. However, very young calves as well as pigs, fowl, and other monogastric animals need to eat foods high in B vitamins. Pigs, poultry, and laboratory animals require specific supplements for the following nutrients, which are most likely to be lacking in regular feeds: riboflavin, niacin, pantothenic acid, and vitamin B12. Poultry feeds may also be lacking in choline. Eggs must need vitamin E to hatch normally. Under some circumstances, it helps prevent muscle stiffness and paralysis (dystrophy) in lambs, calves, and chicks, together with selenium. The majority of other animals can synthesize vitamin C in their bodies and do not require it to be present in their food, as it prevents scurvy in humans and guinea pigs. In cattle that can consume excrement, vitamin K is synthesized by intestinal tract bacteria and is absorbable, therefore dietary supply is typically not a concern. However, as many animals reared today are grown away from feces, vitamin K is frequently given to them in their meals as an added precaution.

### 1.2.5. Antibiotics and Other Growth Stimulants

Since the early 1950s, antibiotics have been a part of livestock diets. These and other non-nutritive compounds known as growth stimulants are given to animal feeds to alleviate illnesses, increase feed acceptance and utilization efficiency, or in some other manner improve the animal's health or metabolism. Antibiotic use can be broadly classified into two categories: therapeutic and subtherapeutic, with the difference between the two based only on the quantity of the drug administered to the feed. In subtherapeutic usage, antibiotics are administered in relatively low dosages to improve animal performance, usually growth and feed efficiency. In therapeutic use, antibiotics are used in sufficient amounts to control bacterial infections within an individual or animal community. When subtherapeutic antibiotics are added to young pig diets, their growth performance is improved by at least 10% to 15%. Antibiotic-resistant bacteria are eventually created as a result of the subtherapeutic use of antibiotics, which makes the development of novel antibiotics necessary for both human and animal use. The two significant antibiotics for human usage, tetracyclines, and penicillin, are being used subtherapeutically in some European Union countries, and there is increasing pressure in the US to do the same. The most often utilized antibiotics in feed are tylosin, penicillin, bacitracin, oxytetracycline, and chlortetracycline. Other growth enhancers that are fed to the animal alter its metabolism rather than preventing sickness. The majority have to do with animal-produced hormones. These include ractopamine for pigs, melengestrol acetate and monensin for cattle, and lasalocid for sheep and cattle. In the US, many ear implants are authorized for the delivery of hormones or other medications to feedlot cattle. For instance, these items usually result in a 10–15% increase in daily gain and a 5–10% increase in feed efficiency. Due to objections from certain consumer groups, other nations, especially those in the European Union, severely restrict or outright ban the use of implants in animals raised for meat.

## 1.3. VALUATION OF FEEDS

The amount of dry matter, protein (including the amino acid composition), fat, fiber, minerals, and vitamins included in the feed can all be found using standard chemical analysis. Complete feed composition charts comprise several energy values (digestible, metabolizable, and net) of the feed, which vary depending on the kind of animal being fed.

### 1.3.1. Determination

Experiments on digestion and balancing gauge how well an animal's body absorbs and retains the different meal ingredients. Amino acids are the building blocks of proteins and are utilized by the animal to grow muscle and other body tissues in different amounts. Cattle and sheep's rumens include microbes that can synthesize amino acids from the many sources of nitrogen in their meal. Pigs and poultry, on the other hand, are monogastric animals and need their food to include the right proportions of vital amino acids. Therefore, monogastrics have requirements for amino acids, but ruminants usually have simple protein requirements. The amounts of digestible protein or amino acids required for growth or other bodily activities are described as protein and amino acid requirements. These numbers can be expressed as a percentage of the diet or as the total grams or units needed per day. Digestion-efficient energy (DE), metabolizable energy (ME), net energy (NE), or total digestible nutrients (TDN) are the units of measurement for energy requirements. The values vary depending on the species. The quantity of heat released when a feed is burned in a bomb calorimeter is its gross energy (GE) value. Using this value has the disadvantage that, because some or all of it travels through the body without being digested, a substance like wood and maize may have a similar GE but very different nutritional qualities. Moreover, urea, a byproduct of digestion, is expelled in the urine. Additional energy is lost as gas produced by bacteria in the digestive tract and from the heat of digestion. Compared to pigs, chickens, and other monogastric animals, this loss is noticeably higher in ruminants. The food energy may also be reduced by the labor required to eat, digest, and metabolize it. This yields the feed's net energy value (NE value), which can be utilized for either production (growth, reproduction, or milk) or maintenance (base metabolism, activity, and warmth). The total digestible protein, digestible ether extract (fat) multiplied by 2.25, digestible nitrogen-free extract (carbohydrate), and digestible crude fiber make up a feed's TDN value. Because it doesn't account for heat losses and fermentation during digestion and metabolism, as well as the fact that energy is used more effectively for maintenance or milk production than for growth and fattening, its use as a measurement factor has decreased recently (Figure 1.1).

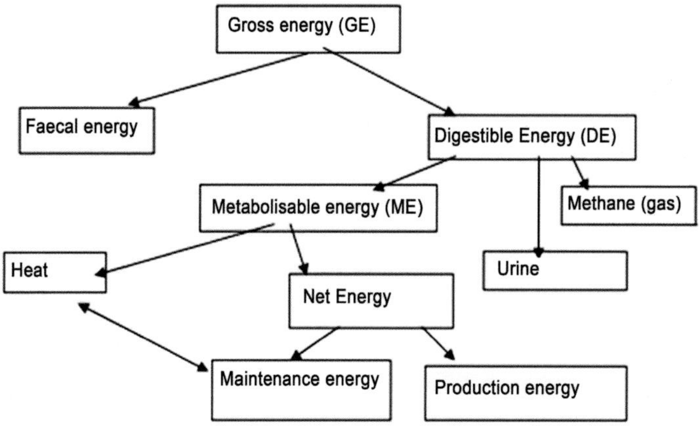


Figure 1.1. *Fate of feed energy within the animal.*

*Source: Mutwedu, V. B., Manyawu, G. J., Lukuyu, M. N., & Bacigale, S. (2020). Fodder production manual for extension staff and farmers in South Kivu and Tanganyika Provinces of the Democratic Republic of Congo. ILRI Manual 37. Nairobi, Kenya: International Livestock Research Institute (ILRI).*

1. **Gross Energy (GE):** The chemical elements that make up the feed can be generically categorized as vitamins, lipids, proteins, and carbs. When the organic material in such feed burns, heat is generated. Because of this, techniques have been devised to determine the amount of chemical energy contained in a feed by calculating the heat produced when a known quantity is burned completely. We refer to this as gross energy. The energy content of the majority of typical meals is roughly 18.5 MJ/kg DM.
2. **Digestible Energy (DE):** The animal can't use all of the gross energy in the diet it has consumed. A portion of the animal's energy is lost through excretions; it is stuck in the feed so the animal cannot get to it. Fecal energy is subtracted from gross energy to determine the digestible energy. The DE is a representation of the nutrient energy content after digestion. The Total Digestible Nutrients (TDN) can be computed from these nutrients that are metabolized.
3. **Metabolizable Energy (ME):** Through gas formation during metabolic activities and urine excretion, the animal further loses energy-containing compounds. After deducting the energy lost from urine and the combustible gases produced by the digestible energy of a feed, what is left is known as metabolic energy. Methane, a

flammable greenhouse gas, can cause significant energy loss, especially in ruminant populations. As a result, there may be detrimental effects on the environment and nutrition.

4. **Heat Increment (HI):** An animal that has consumed feed will also experience energy losses in the form of heat, gases, and chemical energy excretions. Animals constantly generate heat and release it into the environment, either directly via radiation, convection, and conduction or indirectly through the evaporation of bodily water. The nutrients from the meal are broken down and metabolically processed to produce heat. For example, the process of eating involves using muscles to chew, swallow, and secrete saliva, all of which produce heat. This heat energy is useless to the animal unless it lives in an extremely cold climate, thus it must be viewed as a tax on the energy of the feed, much like the energy in excrement. Heat increment is the term used to describe the energy lost in this way.
5. **Net Energy (NE):** A feed's net energy, or the energy an animal can utilize for productive activities including body upkeep and other types of production (milk, meat, eggs, wool, and labor), is obtained by deducting the feed's HI from its ME.

### 1.3.2. Optimization of Nutrient-Cost Ratio

Seasonal variations in feed costs can result in significant savings for farmers if they choose carefully which feed materials to employ in creating comprehensive diets. Large commercial feed corporations with extensive operations find it far simpler to profit from regional fluctuations in feed costs than do individual, relatively small-scale livestock producers who are dependent on feeds provided locally. Computers may be used to determine the precise amounts of competitively priced feed ingredients that together will fully satisfy the nutrient requirements of a certain type of animal at a given stage of development thanks to the least-cost formulation of feed combinations. Computer programs may effectively create diets that produce the most at the lowest cost when utilized by a skilled nutritionist.

## 1.4. ANIMAL FEED MARKET

A vital component of the world's food production, the animal feed market boosts productivity and efficiency in the cattle and poultry industries. The following significant statistics illustrate its importance:

1. **Market Size:** The market for animal feed was estimated to be worth \$460 billion in 2020 and is expected to increase at a compound

annual growth rate (CAGR) of about 5% to reach over $570 billion by 2026.

2. **Regional Distribution**: With almost 40% of the global market share, Asia-Pacific leads the animal feed industry, which is followed by North America and Europe.

Several factors contribute to the growth and evolution of the animal feed market:

- **Rising Demand:** As the world's population is predicted to reach 9 billion people by the year 2050, there will be an increasing need, especially in developing nations, for diets high in protein.
- **Technological Advancements:** Innovation and efficiency in feed production are being propelled by developments in feed formulation technology, such as precision nutrition and the utilization of substitute protein sources.
- **Regulatory Environment:** Strict laws pertaining to animal welfare, environmental sustainability, and food safety are influencing market dynamics and propelling the use of sustainable feed solutions.

The market for animal feed is made up of many product kinds that individually meet the needs of different animal species and nutritional requirements, such as premixes, feed additives, and compound feed. Prominent firms in the sector, like Cargill, ADM, and Nutreco, have a substantial market share amongst them thanks to their extensive global reach and varied product lines.

## 1.5. BASIC TYPES OF FEEDS

Animal feeds are classified as follows:

- **Concentrates:** High-energy-value concentrates, such as fat, high-protein oil meals or cakes (soybean, canola, cottonseed, peanut [groundnut]), and by-products from processing sugar beets, sugarcane, animals, and fish. Cereal grains and their by-products (barley, corn, oats, rye, wheat).
- **Roughages:** These consist of root crops, straw, hays, silage, pasture grasses, and stover (cornstalks).

### 1.5.1. Concentrate Foods

#### *1.5.1.1. **Cereal Grains and Their By-Products***

Barley, corn, oats, rye, and sorghums are farmed primarily for animal feed in the agricultural practices of northern Europe and North America, though minor amounts are also processed for human consumption. For pigs and poultry, these grains are fed whole or ground, either alone or in combination with other by-products, minerals, and vitamins, or as a sufficient dietary supplement for ruminants and horses. Grain production is seasonal due to variations in temperature, moisture content, or both. During the short growing season, a whole year's supply must be produced. To stop it from sprouting or molding, the grain is dried to 14% moisture or below. After that, it is stored in buildings or containers that are safe from rats and insects. Because crop failures do happen occasionally, it is generally preferable to store more than a year's worth of grains to be used as feed.

### *1.5.1.2. High-Protein Meals*

The main purpose of producing vegetable seeds is to extract oil for human use and industrial applications. Examples of these include soybeans, peanuts (groundnuts), sunflower, canola, cottonseed, coconuts, and flaxseed (linseed). Following the extraction of the oil from these seeds, the leftover material—which could include anywhere between 20 and 50% protein and 5 to less than 1% fat—is sold as animal feed. The woody hulls or shells of peanuts and cottonseed are often removed before processing; if the hulls or shells are left in place, the resulting by-product has a higher fiber content and a noticeably lower protein and calorie content. The amount and accessibility of the amino acids in these meals determine how high or low the protein content is for monogastric patients. All ruminants need are supplies of protein or nitrogen so that the rumen microorganisms can produce amino acids. To provide the protein and amino acids required for effective growth or output, these high-protein feeds are added to low-cost roughages, cereal grains, and other low-protein meals. The selection of a supplement for a given diet is mostly influenced by price and availability.

### *1.5.1.3. By-Products of Sugar Beets and Sugarcane*

The sugar beet industry produces dried beet pulp and beet molasses in sugar mills, as well as beet tops, which are utilized on farms either fresh or ensiled. A byproduct of making cane sugar is cane molasses. All of these are delicious, excellent sources of carbs. Stalk residue, or bagasse from sugarcane, has a very poor feed value, is fibrous, and is difficult to digest. Beets and other roots are farmed for animal feed in Europe. As low-cost byproducts of the citrus juice business, citrus molasses and dried citrus pulp are frequently utilized as premium feeds for sheep and cattle.

### *1.5.1.4. Other By-Product Feeds*

The residues or byproducts from the commercial processing of cereal grains for human consumption make up a significant amount of animal feed. The majority of these by-product feeds, such as wheat bran, wheat middlings, wheat germ meal, and wheat mill feed, are derived from milling wheat. Bakery trash, including stale and leftover bread, rolls, and other pastry products, are crushed and fed to farm animals and pets in certain places. Similar methods are used to collect rice hulls and bran from the mills that refine rice for human use. The production of starch for food and industrial purposes results in the by-products of corn gluten feed, corn gluten meal, and hominy feed. Brewer's yeast, soluble from corn distillers' grains, and brewers' grains are valuable animal feeds that are gathered from the dried byproducts of the fermentation processes used to make beer and distilled spirits. Pineapple bran or pulp and the plant's ensiled leaves are waste products from canneries that process pineapples. The meat and bonemeal feeds, tankage (animal waste left over after rendering fat in an abattoir), meat scraps, blood meal, poultry waste and feather meal are examples of by-products from the abattoirs and meatpacking industries that turn animals into meat. Fish-processing factories provide fish meals in a variety of varieties and quality levels. The mineral elements calcium and phosphorus, as well as at least 50% of premium protein, are commonly found in these animal byproducts. These vital minerals are especially abundant in steam-cooked bonemeal. Dairy industry by-products include dried buttermilk, dried whey, and dried skim milk.

## 1.5.2. Roughages

### *1.5.2.1. Pasture*

For ruminants like cattle, horses, sheep, and goats, pasture grasses and legumes—both naturally occurring and farmed—are the most significant single source of nutrition. They provide the majority of these animals' sustenance during the growing season at a cheaper cost than feed that needs to be harvested, processed, and delivered. For grazing animals, hundreds of different types of grasses, legumes, shrubs, and trees constitute suitable feed. Although the cultivated variety's nutritional worth has been investigated, many of the naturally occurring kinds lack sufficient information.

### *1.5.2.2. Hay*

Hay is made by drying legumes or grasses just before the seed develops and as they become close to reaching their maximum development stage. It has been demonstrated that this stage produces the highest yields of carbs and digestible protein per unit of land area. Usually, the moisture level is lowered to less than 18% to avoid heating, molding, and spoiling while being stored. The protein content of legume hays, such as alfalfa and clovers, is high, whereas the protein content of grasses, including timothy and Sudan grass, varies greatly based on their maturation stage and the amount of nitrogen fertilizer given to them. When there is not enough fresh pasture grass, animals are fed stored hay (Figure 1.2).

Figure 1.2. *Bales of hay.*

*Source: Montanabw, I. CC BY-SA 3.0. https://commons.wikimedia.org/w/index.php?curid=2307401 (Creative Common License).*

### *1.5.2.3. Silage*

In order to retain the moist feed, silage is created by placing immature plants in an airtight storage container and allowing fermentation to produce lactic and acetic acids. Storage options include subsurface trenches and tower silos that are upright. Depending on the type of silage, the initial moisture content of the forage should range from 50 to 70%. Lower moisture content can make it more difficult to get enough packing to keep out air, which can lead to mold or other spoiling. An excess moisture level leads to seepage, which depletes nutrients and produces an unpleasant, highly acidic silage. Compared to dried hay, ensiled fodder can be stored for extended periods of time with less nutrient loss. The type of forage ensiled, and the degree of curing success determine the nutritional content of the silage. Silage is made from corn, sorghums, grasses, and occasionally leguminous forages (Figure 1.3).

Figure 1.3. *Silage.*

*Source: By Grochowski, B. (user VOR94) – Own work, Public Domain. https://commons.wikimedia.org/w/index.php?curid=6650185.*

### *1.5.2.4. Root Crops*

For financial reasons, root crops aren't as commonly used as they formerly were as animal feed. Turnips, rutabagas, cassava, beets (mangels), and occasionally extra potatoes are fed to animals. Root crops generally contain energy, but they also have poor dry matter and protein content when compared to other diets.

### *1.5.2.5. Straw and Hulls*

After the crops of wheat, oats, barley, and rice are harvested, the leftover straw is fed to cattle and other ruminants. The straws have a low protein content and a high fiber content, with low digestion. While straw can help sustain mature animals in situations where other feeds are scarce, its low nutritional value makes it unsuitable as a continuous feed source unless combined with other feeds that provide the necessary amounts of protein, digestible energy, and minerals for development and productivity. When straw is treated with alkali, the cellulose's digestibility is significantly increased, increasing the straw's utility to animals as an energy source. In addition, ruminant diets can benefit from the inclusion of corncobs, cornstalks, rice hulls, and cottonseed hulls as sources of fiber. While the others are valued similarly to straw, rice hulls are worth less.

### 1.5.3. Newer Feed Resources

When nutrients needed for the physiological stages are not properly supplied, animals' potential performance is not always realized. This predicament arises from the lack of high-quality animal feed. Furthermore, there is a conflict between fuel, feed, and food since livestock diets contain both edible human components and feedstock for the manufacture of biofuel. Human-edible feeds, such as cereal grains and soybeans, which are important components of monogastric animal diets, may become scarcer for cattle as the world's population grows. For animal production to be sustainable, the highest possible use of substances in feed that are inedible to humans will be needed. It has been determined that, within certain bounds, a variety of non-traditional feeds are utilized in the diets of several animal species. Recently, certain innovative feed resources have been investigated.

#### *1.5.3.1. Insect Meals*

Since they are naturally found in poultry diets and have a high protein content, a variety of insect meals from various species have recently attracted attention as potential sources of protein for monogastric animals. Since insects may be raised on low-grade biowastes and turn them into high-quality protein sources, insect meals have the potential to be sustainable protein feeds. Studies have shown that, depending on the animal species, insect meal can replace 50–100% of fish or soybean meal due to its promising amino acid content and digestibility. To appropriately balance the amino acids and energy in diets, especially for monogastric species, the accurate assessment of the digestibility of amino acids and the amount of metabolizable energy content of various insect meals is necessary. Insect meals might theoretically be used as a substitute for other high-protein feed ingredients (Te Pas et al., 2021).

#### *1.5.3.2. Microalgae*

Microalgae, which include diatoms, green, blue-green, and golden algae, are significant marine resources with the potential to provide significant amounts of protein, bioactive substances, vitamins, and minerals. Animal uses use about one-third of the world's total microalgae production (Te Pas et al., 2021). Solar energy and waste materials like manure can also be used to generate microalgae. Microalgal protein contents range from 25 to 50% depending on the species (Te Pas et al., 2021). Additionally, when included in an animal's diet, microalgae's abundance in n–3 long-chain polyunsaturated fatty acids enhance the quality of the meat and milk (Madeira et al., 2017). Producing microalgae at a reasonable cost is a challenge.

### 1.5.3.3. Seaweed or Macroalgae

In addition to containing up to 60% polysaccharides, seaweed or macroalgae, such as Rhodophyta (red algae), Phaeophyceae (brown algae), and Chlorophyta (green algae), are also rich in high-value components such as n–3 fatty acids, bioactive chemicals, and colorants. Depending on the species, seaweeds can have a wide variety of nutritional profiles. While red algae have 10–30% crude protein and green algae above 15%, brown macroalgae are rich in minerals and contain 5–13% crude protein (Te Pas et al., 2021). Seaweed production has some of the same issues as microalgae production. Due to their high mineral content and suboptimal essential amino acid composition, most seaweed species cannot be used in animal diets unless they are appropriately balanced (Te Pas et al., 2021). Seaweeds have antimethanogenic effects in the rumen, which may be additional advantageous for the environment in addition to their prospective usage as feed ingredients (Min et al., 2021).

### 1.5.3.4. Distiller's Dried Grains with Soluble (DDGS)

A byproduct of the manufacturing of alcohol and biofuel is distiller's dried grains with soluble (DDGS). Depending on the grain stock used to produce the ethanol, it includes high amounts of lipids and proteins and can be an excellent feed for raising cattle. But unlike the cereal grains that are used to make DDGS, it mostly consists of high concentrations of fiber that is difficult to digest, like cellulose, lignin, and arabinoxylans. However, it might partially substitute more expensive conventional feed ingredients, cutting the cost of feed. Furthermore, it has fermented products that provide animals with probiotics, prebiotics, enzymes, and bioactive metabolites; as a result, it may help to enhance production efficiency (Lalhriatpuii and Patra, 2022).

### 1.5.3.5. Tree Foliage

Particularly in arid and semi-arid regions of the world, tree leaf is a particularly valuable source of fodder for small ruminant animal production. It supplements low-quality forage-based diets with additional proteins and minerals (Patra, 2010). Additionally, tree leaves can be used to enhance ruminal fermentation and reduce greenhouse gas emissions (Huang et al., 2021). Crop, vegetable, and fruit residues as well as food wastes can all be used to some degree in livestock diets, which are typically fed to animals in low-income nations. Food wastes and residues from the fruit and other processing sectors must be properly valued in order to be turned from the linear economy into a circular, sustainable bioeconomy that will also lessen its impact on the environment. Utilizing agro-industrial byproducts in animal nutrition is a viable approach to lessen the competition between feed and food, farm-level diet expenses, and the environmental effects of producing food from animals. Additionally, a variety of industrial and therapeutic compounds found in fruit wastes have the potential to enhance animal health and productivity. Utilizing plant secondary metabolites to enhance ruminal fermentation, ruminant productivity, and health while reducing environmental impacts has garnered attention recently.

## 1.6. FEED SAFETY, HEALTH, AND WELFARE

The first line of defense for food safety in the farm-to-table supply chains is livestock feed. For this reason, using safe feeds is essential to ensuring the safety of human food. When raw ingredients are harvested, manufactured, processed, stored, or transported, feeds may become contaminated with biological, chemical, or physical dangers, or they may contain intrinsic toxicants. Specifically, heavy metals, fungus-produced toxins, and pesticides are commonly found in feed. Heavy metal concentrations, such as those found in feeds and water including cadmium, arsenic, lead, mercury, copper, and chromium, are especially common in industrial, urban, and semi-urban areas. In the end, there is a risk to the public's health because meals derived from animals may contain significant quantities of these heavy metals. As a result, there needs to be extra care taken to minimize heavy metal contamination in these areas and take preventative steps such as adding nutritional supplements to milk and meat to lower their levels. The animal sector frequently adds in-feed antibiotics, but they are worried about the emergence of microorganisms that are resistant to antibiotics, which could endanger human health. Many studies have reported a potential link between the use of antibiotics at subtherapeutic levels and the development of antimicrobial resistance among the microbiota, despite differing opinions regarding the transfer of antibiotic-resistance genes from animal pathogens to human pathogens (Lillehoj et al., 2018).

In recent decades, a number of substitutes have been investigated, such as probiotics, synbiotics, organic acids, phytogenics, enzymes, antimicrobial peptides, bacteriophages, clay, and metals. Compared to antibiotics, these feed additives have superior benefits on immunomodulation, gastrointestinal health, and antioxidant status. While many of the alternatives have shown encouraging results, it is unclear how they work, how effective they are, and what benefits and drawbacks come with using them (Lillehoj et al., 2018). Animal welfare in cattle production systems is compromised by a variety of pressures, including overcrowding, transportation, temperature, and overproduction. Stresses impair gut health, immunity, animal performance, and increase susceptibility to illness in animals. Appropriate feeding techniques, such as the application of medicinal plants, gut microbiota-acting agents, and antioxidant vitamins and minerals, can mitigate a variety of stresses and enhance animal immunity, antioxidant status, and gut health in addition to improving animal productivity and the quality of products stored in storage (Patra and Kar, 2021). Because of their high nutrient intake or high metabolic rate, grill chickens and turkeys that are overgrown are more likely to develop numerous metabolic diseases, the majority of which are related to cardiovascular (such as ascites and pulmonary hypertension syndrome) and musculoskeletal (such as lameness, dyschondroplasia, and spondylolisthesis) disorders. These diseases also result in greater economic loss than infectious diseases. When high-yielding cows are fed large amounts of grains to meet their energy needs, subacute ruminal acidosis frequently results. This lowers milk output, ruminal health, and barrier function.

### 1.6.1. Food Quality

Demand for foods that improve health, and well-being is rising as a result of consumers' growing interest in eating healthily. Through selective feeding, the concentration of various health-promoting fatty acids in milk and meat can be efficiently increased. Decreases in saturated fatty acid content and increases in n–3 fatty acid and rumenic acid (cis-9, trans-11 C18:2) in milk and meat have been the focus of several investigations. With varying degrees of success, a wide range of plant secondary compounds have been studied. These include polyphenols (simple phenolic compounds, tannins, and flavonoids), essential oils, and saponins, which have particular antimicrobial effects in the rumen responsible for fatty acid biohydrogenation (Roy et al., 2013). Tannins and essential oils have been shown in some research to have no therapeutic effects, while in other studies they have been found to be successful in blocking the first stage or, less frequently, the last step of polyunsaturated fatty acid biohydrogenation (Toral et al., 2018). Higher antioxidative plant secondary compounds can improve antioxidant status, lower lipid peroxidation and the deterioration of meat and milk quality during storage and reduce volatile chemicals like skatole and indole—which cause an off-flavor in meat. The causes of inconsistent effects, which could be related to the many active chemicals, ruminant animal species, dose, feed composition, and physiological stages, would require more investigation (Toral et al., 2018).

### 1.6.2. Animal Feed Industry Trends and Challenges

Numerous issues facing the feed business have a direct impact on the world food chain. These include:

- **Increased Demand for Animal Protein:** By 2050, it is expected that there will be 9.7 billion people on the planet, which would raise the demand for animal protein. To meet this demand, the animal feed sector will need to increase output while maintaining sustainability and minimizing the negative environmental effects of raising animals.

- **Technological Advancements:** Technological developments are revolutionizing the animal feed sector. Precision feeding systems, for instance, can assist farmers in maximizing feed utilization and minimizing waste. Additionally, novel feed components that are more environmentally friendly and sustainable are being developed through the application of biotechnology.
- **Shift Towards Sustainable Feed Production:** Customers are calling for more environmentally friendly food options as they grow more conscious of the effects that cattle rearing has on the environment. As a result, there has been a change in the feed production industry towards sustainable methods including using algae, insect protein, and other substitute feed ingredients.
- **Environmental Sustainability:** Deforestation, water pollution, and greenhouse gas emissions are all significantly impacted by the raising of livestock. The animal feed sector needs to figure out how to satisfy the growing demand for animal protein while lessening its negative effects on the environment.
- **Feed Safety and Quality:** For both human health and the wellbeing of livestock, the safety and quality of animal feed are essential. The sector is responsible for guaranteeing that feed is nutrient-rich and free of impurities.
- **Regulatory Compliance:** Many rules and guidelines that apply to the animal feed sector might be confusing and difficult to follow. Adherence to these regulations is crucial in guaranteeing the safety and caliber of feed.
- **Supply Chain Interruptions:** The profitability of the feed business will continue to be impacted by the high costs of raw materials and feed additives brought on by global supply chain constraints. During the peak of the COVID-19 epidemic, supply chains faced significant challenges that affected all industries globally. The agriculture sector had a number of knock-on consequences, including severe disruptions to both local and international supply chains, the hoarding of goods, especially animal feed, shortages of vital feed components, and staff reductions, to mention a few.
- **Fluctuations in Ingredient Sourcing and Pricing:** In animal husbandry, feed is essential since it serves as the main source

of sustenance for the animals. For farmers to raise healthy and productive animals, they must source high-quality feed; but, because of the volume of feed needed per animal and the narrow profit margins in livestock production, this can be a significant investment. Feed prices can significantly exceed other input costs for animals like pigs and poultry, making up as much as 75% of the total cost of production per animal. Feed prices have increased sharply in recent years due to changes in the feed milling business and changes in the market. This has threatened the bottom line of livestock farmers by having a substantial impact on the expenses and profitability of producing cattle. The primary cause of the growing cost of feed is the fluctuations in vital inputs used in feed milling. The process of grinding and processing raw materials into a state that is nutrient-rich and fit for animal consumption is known as feed milling. Three main inputs are needed for this: labor, raw materials, and energy. Regretfully, there are historic shortages in all three. Feed milling is just one of the many supply chain operations where rising operational costs have been mostly attributed to increases in gas and power prices. The cost of raw materials has skyrocketed at the same time as a result of several subpar wheat and maize harvests and the ensuing scarcity of supply. The situation has been made worse by the ongoing Russo-Ukrainian conflict, which has affected the export and regional production of feed ingredients including wheat and soybean as well as sunflower meal, which is produced in Ukraine and accounts for over half of the world's supply. Logistics have also been affected by the conflict; export feed ingredients are now more expensive, more time-consuming to ship, and more likely to spoil. In addition, the feed industry is now directly competing with the quickly expanding bioenergy sector for raw materials, with poorly funded feed mills finding it difficult to survive against the heavily subsidized bioenergy sector.

*Chapter* 2

# Animal Feed Nutrients and Additives

## Contents

# 2.1. CARBOHYDRATES

## 2.1.1. Overview

Approximately 60% to 90% of the dry matter (DM) in plant tissue is composed of carbohydrates, which are the main constituents. Carbohydrates are hydrates of carbon because they have carbon, hydrogen, and oxygen in the same proportion as water ($CH_2O$). In animal cells, carbohydrates serve as the primary energy source. For the animal, dietary carbohydrates derived from plant-based foods are a primary source of energy. Carbohydrates can be found in plant cells as sugar or starch or they can be linked to the structure of the cell wall (e.g., cellulose). Animal cells use metabolic processes to release energy from the carbohydrates in plant materials (such as cereal grains, grass, and fodder) when they consume them. In general, animal metabolism generates energy through a method that is opposite to photosynthesis.

## 2.1.2. Carbohydrates Classification

The number of carbon atoms in each carbohydrate molecule and the total number of sugar molecules in the compound are two factors that can be used to categorize carbohydrates. A carbohydrate is categorized as triose (3 C), tetrose (4 C), pentose (5 C), or hexose (6 C) depending on how many carbon atoms it contains. A biological term that ends in "ose" designates the molecule as a "sugar." The most prevalent sugars in animal tissues are pentoses, such as ribose in ribonucleic acid (RNA), and hexoses, such as glucose or blood sugar. Carbohydrates are categorized into the following groups based on the number of sugar molecules they contain:

### *2.1.2.1. Monosaccharide, One Unit of Sugar*

Since they cannot be hydrolyzed into smaller molecules, monosaccharides—such as glucose—are commonly referred to as simple sugars. Monosaccharides can be classified by the number of carbon (C) atoms they contain. The prefixes indicating the number of carbons in a sugar molecule are listed below:

- Triose (3 C);
- Tetrose (4 C);
- Pentose (5 C; e.g., xylose and ribose);
- Hexose (6 C; e.g., glucose, fructose, galactose, and mannose).

In animal tissues, 5 C and 6 C sugars make up the majority of monosaccharides. Aldose, some sugar with an aldehyde structure, and ketose, some sugar with a ketone group, are further divisions of simple sugars. Having the same chemical formula ($C_6H_{12}O_6$), glucose and fructose are both hexoses (6 C). However, fructose is a ketose, or a ketohexose, and glucose is an aldose, also known as aldohexose. Glucose, fructose, and galactose are the three hexoses that are crucial for nutrition and metabolism. Glucose's chemical structure can be shown in both cyclic and straight chain forms. Glucose is mainly found in cyclic form (in aqueous solution) and is rarely found in a straight form in biological systems. The main type of carbohydrate that the body uses to produce energy is glucose, which is the form of carbohydrates found in blood sugar. Dried fruits and honey include fructose, sometimes known as "fruit sugar," which is produced when the disaccharide sucrose is broken down. Mammalian milk contains the disaccharide lactose along with galactose, which is produced after digestion. Glucose has significant nutritional significance for animals and can exist as α and β isomers. The direction of the OH on C #1 varies between these two isomers (Figure 2.1).

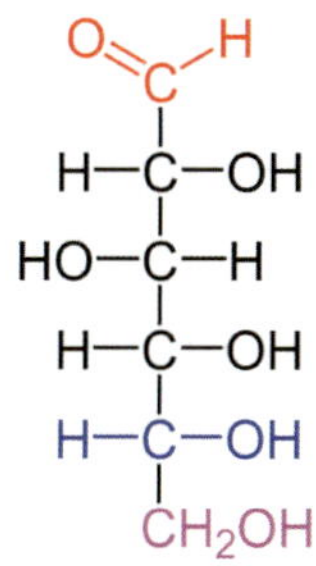


Figure 2.1. *Carbohydrates: An example is D-glucose.*

*Source: By ClockworkSoul at English Wikipedia – Transferred from en.wikipedia to Commons. Public Domain, https://commons.wikimedia.org/w/index.php?curid=3534327.*

For instance, stiff polymers with β-D-Glucose are found in cellulose, but starch has α-D-Glucose. Sugars that are necessary for nutrition are in the D-form, not the L-form. The stereo-orientation at asymmetric carbon position 5 in a hexose or carbon position 4 in a pentose is denoted by the letters D and L.

### *2.1.2.2. Disaccharide, Two Monosaccharides*

Two monosaccharides joined by a glycosidic (covalent) bond form disaccharide. Some examples of common disaccharides are the following:

- Sucrose-glucose + fructose (e.g., table sugar);
- Lactose-glucose + galactose (milk sugar);
- Maltose-α-D-glucose + β-D-glucose (malt sugar);

- Cellobiose-β-D-glucose + β-D-glucose (cellulose).

The only other disaccharide that is derived from animals is lactose, sometimes known as milk sugar. Cellobiose, a part of cellulose, is crucial for animal nutrition, nevertheless. Because monogastric animals lack the cellulase enzyme necessary to break β-D-glucose, they are unable to digest cellulose (Figure 2.2).

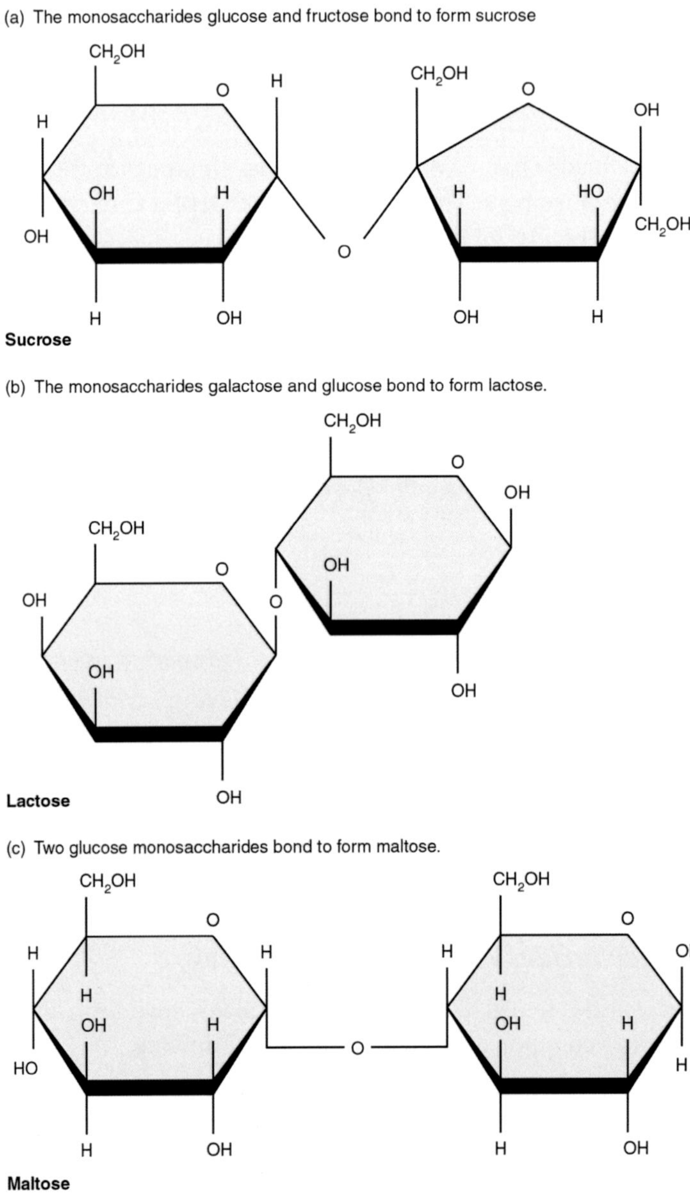

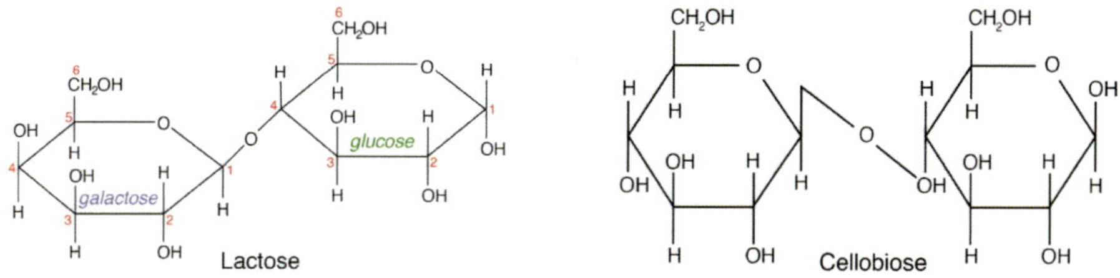


Figure 2.2. *Important disaccharides in animal nutrition and feeding.*

*Source: https://commons.wikimedia.org/wiki/File:218_Three_Important_Disaccharides-01.jpg.*

### *2.1.2.3. Oligosaccharide, 3 to 15 Monosaccharides*

Three or more (3–15) monosaccharides bound together can be combined to form an oligosaccharide.

- Raffinose (glucose + fructose + galactose; 3 sugars);
- Stachyose (glucose + fructose + 2 galactose; 4 sugars).

In animal diets, legumes and beans are frequently sources of oligosaccharides. Certain oligosaccharides are employed as prebiotics—substances that promote the growth of beneficial bacteria. The usage of various oligosaccharides (such as fructooligosaccharides and mannan oligosaccharides) as feed additives to improve hindgut health has garnered more attention recently.

### *2.1.2.4. Polysaccharides, Large Polymers of Simple Sugars*

As their name suggests, polysaccharides are formed by combining vast polymers of simple sugars. The most significant type of carbohydrates in animal feed are polysaccharides. Numerous individual monosaccharide molecules joined in lengthy, intricate chains make up polysaccharides. Polysaccharides serve a variety of purposes in biology, such as energy storage (seed starch in cereal grains) or structural support (glycogen in animal cells) (plant fiber). In animal nutrition, components of cell wall composition are also known as non-starch polysaccharides, or resistant starch, because they are fermented by rumen and hindgut bacteria rather than being absorbed by animal enzymes. It is possible for polysaccharides to be either homopoly or heteropolymeric.

### *2.1.2.5. Homopolysaccharide*

There is only one kind of saccharide unit found in homopolysaccharide. In animal nutrition, homopolysaccharides such as cellulose (the structural form of plants), starch (the nonstructural form), and glycogen (the animal form) are important.

1. **Starch:** The primary carbohydrate in cereal grains is sucrose (seed energy storage). Alpha D-Glucose is the fundamental unit. Among the starches found in cereal grains are:
   - **Amylose-α 1,4 linkage-straight chain, nonbranching, helical structure:** The most basic polysaccharide is amylose, which is made up entirely of glucose units connected by an alpha 1,4 bond. In most plants, amylose—which is soluble in water—makes up 15% to 30% of total starch.
   - **Amylopectin-α 1,4 linkage with alpha 1,6 linkage at branch points:** The way the glucose units are bonded together varies in amylopectin. Although alpha 1,4 bonds are more common, an alpha 1,6 linkage can give rise to a "branch." The structure of amylopectin is more complex than that of amylose because of these branches. In plant cells, amylopectin makes up between 70% and 85% of the total starch content. It is not soluble in water.
2. **Glycogen:** Animal starch refers to a type of starch called glycogen that is present in animal tissue. Physically similar to amylopectin, glycogen is a polysaccharide consisting of basic alpha-D-glucose with a mixture of α 1,4 and α 1,6 linkages. There is very little glycogen (less than 1%) in muscle and liver tissue.
3. **Cellulose:** The most prevalent type of carbohydrate in nature is cellulose. Plant cell walls are given structural integrity by it. β 1,4 linkage, straight chain, nonbranching is the fundamental unit. Very stable is cellulose. Only microbial cellulase has the ability to break it down; no vertebrate enzyme can do so. However, the rumen of ruminant animals, including cattle, contains bacteria that produce the enzyme cellulase. To release the sugar for energy, it destroys the beta 1,4 connections between the glucose molecules in cellulose.

#### *2.1.2.6. Heteropolysaccharide*

Plant cell walls include a mixture of 5 C, and 6 C sugars called heteropolysaccharides (hemicellulose and pectin, for example, which are a combination of pentose and hexose units).

### 2.1.3. Digestion and Absorption of Carbohydrates

The small intestine's lumen is the main location of carbohydrate digestion, where pancreatic amylase starts to break down starch granules (amylose and amylopectin). Farm animals do not have salivary amylase activity in their mouths, while certain birds do. Amylase comes in two varieties: one breaks down α 1,4 links randomly, while the other eliminates disaccharide units (maltose) from the polysaccharide chain. The α 1,6 bonds that constitute the branch points in amylopectin's structure are not acted upon by pancreatic amylase. After being broken down by amylase, a combination of glucose, maltose, and dextrins (residues with α 1,6 branch points) are produced. α 1,6 glucosidase acts on dextrins. Dietary simple sugars like fructose and glucose can be readily absorbed via the intestinal epithelium, negating the requirement for digestion. The final stages of digestion take place at the brush boundary, where the products of starch digestion disseminate. Enzymes on the intestinal brush border hydrolyze disaccharides like maltase and isomaltase to release monosaccharides, which are subsequently absorbed by the enterocyte. This process of degradation continues until the monosaccharides are released. When sucrase reacts with sucrose, it produces glucose and fructose that can be absorbed. Lactase breaks down lactose in young animals raised on milk (preweaning) to produce glucose and galactose. When young animals drink milk, their amylase activity is very low and is boosted by the absorption of solid food. Both simple diffusion and adenosine triphosphate (ATP)-dependent active transport are capable of absorbing monosaccharides. Binding both glucose and Na+, a sodium-dependent glucose transport protein moves the materials through the enterocyte and releases them into the cytosol (Table 2.1).

Table 2.1. Overview of Carbohydrate Digestion

| **Site** | **Enzyme** | **Product** |
|---|---|---|
| Mouth | Salivary amylase | Not too significant |
| Small intestine | Pancreatic amylase di/ oligosaccharidases | Monosaccharides |
| Large intestine | None | Some volatile fatty acid |

*Source: https://open.oregonstate.education/animalnutrition/chapter/chapter-4/.*

Animals with monogastric digestion lack the enzymes necessary to break down complex carbohydrates (β 1,4 linkages), such as glucans, cellulose, and nonstarch polysaccharides (NSP), which are found in plant fibers like wheat and barley and are broken down by bacteria in the hindgut to produce volatile fatty acids (VFAs). In a monogastric diet, high amounts of NSP and glucans might generate viscous digesta, which can impede digestion and result in malabsorption. Feeds high in non-starchy petroleum (NSP) (such as barley and rye) can cause diarrhea, moist litter, and unclean eggs in poultry.

## 2.1.3.1. Carbohydrate Digestion in Ruminants

In ruminant animals, the process of digesting carbohydrates involves microbial fermentation within the rumen. Dietary carbohydrates are broken down (fermented) by bacteria, fungi, and protozoa found in the rumen. The goal of rumen fermentation is to provide the bacteria with ATP, which they can use for protein synthesis and self-growth. VFAs, often referred to as short-chain fatty acids (see below), are created as a byproduct of fermentation in the rumen. The animal absorbs them through the rumen wall and uses them as fuel. Acetic (C2), propionic (C3), and butyric acid (C4) are the three main volatile fats (VFAs). The reticulum and rumen are relatively tiny and underdeveloped in juvenile ruminants. A tube-like fold of tissue called the reticular/esophageal groove reflex directs milk or water sucked from a nipple through the omasum and into the abomasum. This is a reflex that sucking triggers. The animal typically loses this instinct after being weaned. When solid food, like creep feed, enters the tiny rumen, fermentation begins. Although the neonatal ruminant animal does not yet have a ruminal bacterial population, it begins absorbing germs from its mother and surroundings as soon as it is born, especially through touch. After solid food is fermented, volatile fatty acids (VFAs) are produced. These acids promote the rumen's growth and development, especially the papillae's growth for absorption. Microbial cell masses, or VFA produced by microbial protein synthesis, and gases such as carbon dioxide, methane, hydrogen, and hydrogen sulfide are the final byproducts of rumen fermentation. The relative makeup of the rumen bacteria will affect the fermentation products. Because diet alters the substrates for fermentation and, in turn, the products of fermentation, it also affects the microbial population. For instance, the main nutritional component of ruminants fed concentrate (feedlot cattle) is starch. These rumens will contain more amylolytic bacteria than the rumens of animals fed pasture and roughage, which have cellulolytic bacteria. The rumen fermentation process and the formation of volatile feed additives can be influenced by various factors, including the forage:concentrate ratio, the physical shape of the diet (ground vs. pelleted), feed additives, and animal species. The diet's forage:concentrate ratio affects the molar ratios of VFAs. Amylolytic bacteria tend to create more propionic acid, whereas cellulolytic bacteria prefer to produce more acetate. Three main VFA molar ratios are typically 65:25:10 for a diet high in roughage and 50:40:10 for a diet high in concentrates. Variations in volatile fat content (VFA) can cause a number of digestive problems in ruminants. When cattle are rapidly moved from pasture or range feeding to feedlot conditions, or when their diets are high in grains, rumen acidosis develops in the animals.

### *2.1.3.2. Carbohydrates, Metabolism*

"Ability to do work" is the definition of energy. All of the bodily functions that are necessary for an animal to survive, such as the transportation of nutrients, synthesis, and contraction of muscles, require energy. An animal cannot move, consume its food, develop, reproduce, or even breathe without energy. Since food-producing animals must synthesize nutrients (such as proteins and fat) for deposition into muscle, milk, and eggs, energy requirements and balance are particularly crucial. The main source of energy in farm animals' diets is carbohydrates.

### *2.1.3.3. Energy Forms*

An important part of energy metabolism is hydrogen. Nicotinamide adenine dinucleotide (NAD+) and flavin adenine dinucleotide (FAD) are two examples of hydrogen receptors that receive hydrogen from glucose during the animal's breakdown of glucose ($C_6H_{12}O_6$). To liberate energy, these hydrogen acceptors, also known as reducing equivalents, are oxidized during the respiratory chain processes inside the mitochondria. In biological systems, the production of adenosine triphosphate (ATP) is correlated with the oxidation of hydrogen. The cell's easily accessible energy source, or "molecular energy currency unit," is ATP. Three substances make up ATP: triphosphate, ribose, and adenine, a nitrogenous base. The $PO_4$ bonds store energy, and each phosphate bond break releases eight kcal of energy.

### *2.1.3.4. Glucose Utilization*

The animal may employ the glucose it has ingested for a variety of purposes. The production of glycogen in the liver, which is stored in hepatic and muscular tissue, is the primary priority. Nevertheless, the body doesn't store much glycogen, a substance that resembles starch, and glycogenolysis is a quick process that converts glycogen back into glucose. In healthy, typical animals, this aids in keeping blood glucose levels within a specific range. Oxidation, a primary function that produces energy, is the second priority and is covered in the following sections. Excess glucose is converted to fat and stored in adipose tissue because the amount of glycogen that can be stored is restricted. This is achieved by the process known as glycolysis, which breaks down glucose to pyruvate, which is subsequently used for the creation of fat. Different enzymatic activities are used by cells in different steps to convert the ingested glucose to carbon dioxide and water. Two metabolic pathways—glycolysis and the tricarboxylic acid (TCA), commonly known as citric acid or Kreb's cycle—are involved in the catabolism of glucose.

### 2.1.3.5. Glycolysis

Glycolysis takes place in the cytosol of the cell, which contains the enzymes necessary for the process. In aerobic metabolism, glycolysis breaks down glucose (6 C) into two 3 C end products, pyruvates, and in anaerobic metabolism, lactic acid. It is an oxidative catabolic process that produces reduced NAD (NADH) and ATP. Other sugars, such as fructose and galactose, are catabolized by the process of glycolysis, which involves turning them into glycolysis intermediates. Hexokinase has the ability to convert fructose to fructose-6-phosphate. Galactose can enter glycolysis through conversion to galactose-1-phosphate, which can next be transformed to glucose-1-phosphate and, finally, to glucose-6-phosphate (G6P), an intermediary in glycolysis. Two phases make up glycolysis:

- an energy investment phase requiring the input of ATP (preparatory phase); and
- an energy realization phase (pay off) where ATP is made.

Hexokinases, an enzyme found in glucose-consuming cells, phosphorylate glucose—that is, attach a phosphorus group—and convert it into G6P using ATP. The "machinery" inside the cells can now start to metabolize the glucose. In short, hexokinase catalyzes the transfer of phosphate from ATP to glucose in the first step of glycolysis, resulting in the formation of glucose-6-phosphate. As a result, ATP is used in this stage, providing the energy required for the reaction to continue. After being transformed from glucose-6 phosphate to fructose-6-phosphate and then fructose-1,6-biphosphate, dihydroxy acetone phosphate (DHAP) and glyceraldehyde-3-phosphate (G3P) are produced by cleaving the latter. An extra ATP is needed for this action to phosphorylate the intermediate fructose-6-phosphate. As a result, for every glucose molecule metabolized during the "preparation," two molecules of ATP are required. Pyruvate is created by further processing G3P during the payout phase. In the intermediate steps of this phase, one NADH and two ATP are generated. The generated DHAP is easily translated into G3P and handled similarly to the initial G3P. Thus, two NADH, four ATP, and two pyruvate molecules will be produced for every glucose molecule (Figure 2.3).

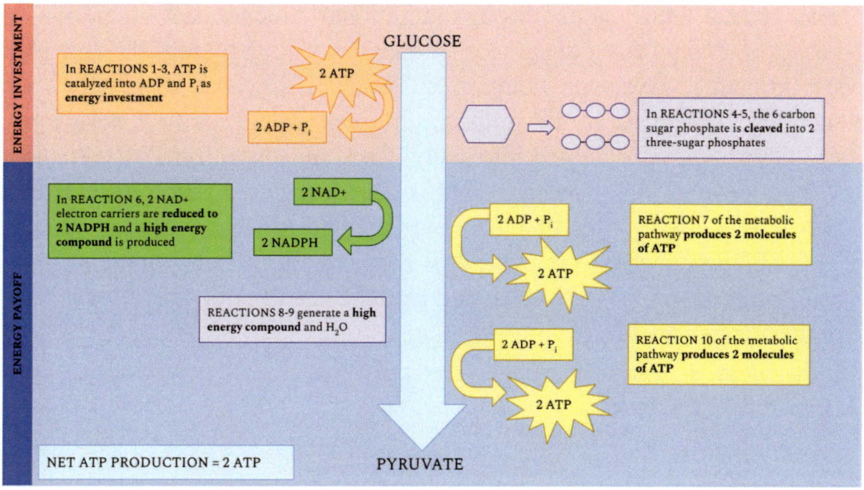


Figure 2.3. *Glycolysis pathway in cytosol.*

*Source: By Loh, S. – Own work, CC BY-SA 4.0, CC BY-SA 4.0. https://commons.wikimedia.org/w/index.php?curid=118301216.*

It's critical to talk about what happens to the pyruvate produced during glycolysis. The fate of pyruvate varies based on the animal's circumstances and the type of cell.

- **Lactic Acid Production:** Aerobic cells convert pyruvate to acetyl coenzyme A (CoA) for oxidation in the citric acid cycle because there is an abundance of NAD+ when oxygen is present. NAD+ levels can drop in the absence of oxygen; therefore, lactate dehydrogenase uses NADH to convert pyruvate to either ethanol (bacteria/yeast) or lactate (animals). Cells receive far less ATP energy from the anaerobic conversion of NADH to NAD+ than they do from the presence of oxygen. Just two ATP are produced by the anaerobic metabolism of glucose. Another mechanism will turn the lactic acid back into pyruvate and make glucose if the cell's supply of oxygen is exhausted.
- **Acetyl CoA Production:** As the primary precursor for the TCA cycle, lipogenesis, and ketogenesis (when negative balance), acetyl CoA synthesis happens in the aerobic state. The TCA cycle has several stages that convert acetyl CoA to ATP. Through a series of condensation, isomerization, and dehydrogenation reactions, the enzyme pyruvate dehydrogenase and various coenzymes containing B vitamins (thiamine, riboflavin, niacin, and pantothenic acid) function during this conversion and produce several different intermediates that are used for the synthesis of fat or amino acids.

The animal body goes through additional biochemical operations to produce extra energy from the glucose molecule. Among these is the enzymatic process known as pyruvate dehydrogenase (PDH), which serves as a link between the TCA cycle in the mitochondria and glycolysis (cytosol). In this stage, 3 C pyruvate is changed into acetyl CoA, an active form of acetic acid, and $CO_2$ is created. After pyruvic acid is decarboxylated and NAD+ absorbs the two H ions, two moles of NADH are created for every mole of glucose (net = 6 ATP produced). Coenzyme A is required for this enzymatic step, and the concentrations of ATP, NADH, and acetyl CoA all strongly influence how active the step is.

### *2.1.3.6. Kreb's Cycle*

The TCA cycle, also known as the citric acid cycle or Kreb's cycle (named after Hans Krebs, who discovered the mechanism in 1937) is a sequence of chemical events that take place in the mitochondrial matrix and are catalyzed by enzymes. Thiamine, riboflavin, and niacin are a few examples of the coenzymes in the pathway that are present in B vitamins.

The essential component of aerobic respiration in cells is the TCA cycle. In the animal body, the TCA cycle provides precursors for the storage of fuels (lipids) and building blocks (amino acids). As previously mentioned, pyruvic acid undergoes PDH conversion to acetyl CoA before entering the cycle. To join the cycle, acetyl CoA (2 C) must combine with oxaloacetate (4 C) to create citric acid (6 C). The cycle is "pulled" onward by this response. About 10 different enzyme-catalyzed transformations of citric acid result in diverse intermediates, including fumarate, malate, succinate, and α-ketoglutarate. High-energy electrons are released to NAD in several of these processes, and NAD molecules also pick up H+ to produce NADH. FAD acts as the electron acceptor in one of the stages, absorbing H+ to create FADH2. One acetyl CoA is also used to create one ATP in one of the processes.

The end result of the cycle is oxaloacetic acid, which is the same as the acid at the start of the cycle and will take up another acetyl CoA to start the cycle again. Six NADH, two FADH2, two ATP molecules, and one mole of glucose or two moles of pyruvic acid are produced overall by the TCA cycle. The electron transport chain uses both FADH and NADH to produce ATP. The TCA cycle yields a net gain of 24 ATP per mole of glucose overall (Figure 2.4).

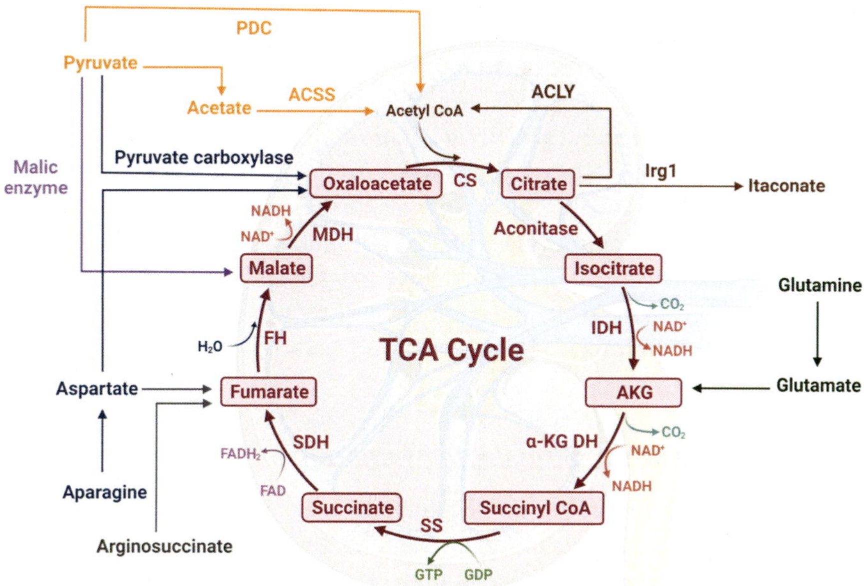


Figure 2.4. *TCA cycle.*

*Source: Jiménez-Uribe, A. P., Hernández-Cruz, E. Y., Ramírez-Magaña, K. J., & Pedraza-Chaverri, J. (2021). Involvement of tricarboxylic acid cycle metabolites in kidney diseases. Biomolecules, 11(9), 1259. https://doi.org/10.3390/biom11091259.*

In total, we get 38 ATP for each glucose molecule that is completely broken down into $CO_2$ and $H_2O$. Every ATP high-energy phosphate bond contains eight kcal of energy. Therefore, the energy recovered net is $38 \times 8 = 304$ kcal. Thus, $304/674^{*}1 \times 100 = 45\%$ is the efficiency of transforming the energy of the glucose bond into an ATP high-energy P bond (*=gross energy of 1 mole of glucose). The reaction of oxaloacetate with acetyl CoA to form citrate is the TCA cycle's rate-limiting phase. For this reason, to generate ATP and preserve cell viability, sufficient amounts of oxaloacetate are required. Nutritional disorders like ketosis, which arise from insufficient energy generation, necessitate the use of oxaloacetate as a precursor to initiate the TCA cycle. In addition, the TCA cycle yields intermediate that act as a starting point for the synthesis of amino and fatty acids. For instance, oxaloacetate can be utilized for nonessential amino acids (glucogenic amino acids or most of the nonessential amino acids, such as alanine, serine, cysteine, and glycine) and citrate for the synthesis of fatty acids. These amino acids can create glucose under low glucose conditions, and they can also be used to synthesize glutamic acid by α-ketoglutarate.

The electron transport chain enables NADH and FADH-related redox energy to be recovered. The electron transport system provides metabolic water by reacting H ions with $O_2$ to create water. NADH and FADH2 transfer electrons to the mitochondrial wall's electron transport system. Coenzyme Q serves as a "traffic cop" for electrons by accepting a pair of electrons and sending them to cytochrome C separately. Since oxygen serves as the terminal electron acceptor, electrons cannot go through the electron transport system and re-oxidize NADH and FADH2. The citric acid cycle will also not function due to these causes. A portion of metabolic control is this. Reactive oxygen species (free radicals) can be produced when there is an interruption in the flow of electrons. Reactive oxygen species are deactivated with the aid of cellular enzymes like catalase and superoxide dismutase.

### *2.1.3.7. Ruminant Carbohydrate Metabolism*

More than 70% of the energy used by ruminants comes from volatile fatty acids (VFAs), which are produced in huge quantities during ruminal fermentation. Through diffusion over a concentration gradient, the rumen epithelium efficiently absorbs VFAs. Each of the many VFAs experiences a distinct level of metabolism as it moves through the epithelium. While practically all of the butyric acid is converted into beta-hydroxybutyric acid, a kind of ketone body, in the epithelium, acetate and propionate pass through it essentially undisturbed. The three main VFAs that are absorbed from the rumen—acetic, propionic, and butyric—have relatively different metabolic outcomes: The majority of the body oxidizes acetic acid, which the liver uses very little of to produce ATP. Acetate is also essential for the TCA cycle because it is the primary source of acetyl CoA. Lipids (like milk or body fat) are synthesized using acetic acid. Acetic acid synthesis is favored by a diet high in roughage. The liver virtually totally excises propionic acid from portal blood. After being transformed into succinyl CoA, propionate joins the TCA cycle. Propionate is a significant substrate for gluconeogenesis in the liver, which is vital to ruminants because very little glucose is absorbed in the small intestine. For instance, the liver of a dairy cow synthesizes all of the glucose in the milk lactose, with propionic acid accounting for the majority of this synthesis. Propionic acid synthesis is favored by a diet heavy in concentrates. Butyrate enters the TCA cycle after being divided into two acetyl CoA molecules by butyric acid. The majority of butyric acid that exits the rumen is converted to the ketone beta-hydroxybutyric acid, which is oxidized in various tissues to produce energy.

## 2.2. Lipids

Lipids, sometimes referred to as fats, are found in both plant and animal tissues, such as vegetable oils, meat, eggs, and milk. Lipids have the physical property of being soluble in organic solvents like hexane, ether, and chloroform, but they are highly insoluble in water. Chemically, lipids are organic substances that are esters of fatty acids, glycerol (a 3-carbon molecule), and other alcohols. Lipids can be categorized as:

- Simple lipids – esters of fatty acids with alcohols. For example, 1 glycerol + 3 fatty acids form a triglyceride (also known as triacylglycerol).
- Compound lipids – glycolipids, lipoproteins, and phospholipids.
- Derived lipids.

Triglycerides, a type of simple lipid that is more prevalent, are a crucial part of animal diets, along with vegetable oil and animal fats like lard or tallow. Proteins, for example, are nonlipid molecules combined with lipids to form compound lipids. Compound lipids, such as lipoproteins (lipid + protein), are utilized for lipid transport, acting as a sort of messenger. Compound lipids play a larger role in physiology and metabolism within the animal body (phospholipids in cell membranes, lipid transport, etc.). As their name suggests, derived lipids are produced via hydrolytic reactions from simple or compound lipids. Sterols, fatty acids, and fat-soluble vitamins are derived lipids. From a nutritional standpoint, lipids are vital for animal existence and are great sources of energy. For animals, fats are the only supply of necessary fatty acids (those that the body is unable to produce). Additionally, fats can supply fat-soluble vitamins. But when vitamins are added to cattle feed, this function is actually quite little. From a physical point of view, adding fats improves feed quality, lowers dust in feed, reduces feed particle separation during processing, increases palatability, lubricates the digestive tract more effectively (emulsification and passage rate), and increases feed digestibility. The primary components of lipid nutrition are fatty acids. Their differences in metabolizability, composition, and structure are the cause of this. A fatty acid is made up of two groups at opposite ends of a hydrocarbon backbone: a hydrophobic methyl group (–$CH_3$) and a hydrophilic carboxyl group (–COOH) (Figure 2.5).

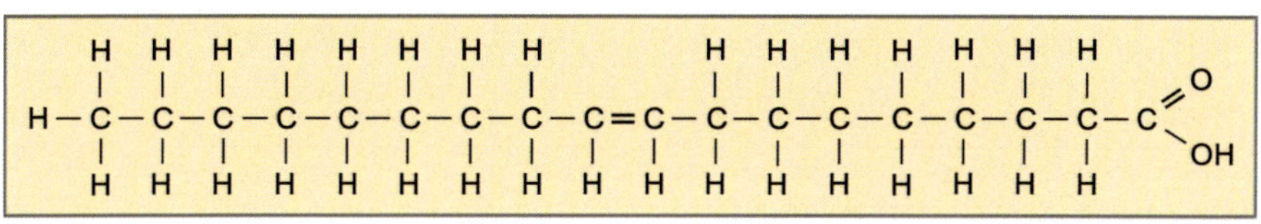

Figure 2.5. *Fatty acid structure with hydrocarbon chain showing carboxyl and methyl end.*

*Source: https://courses.lumenlearning.com/suny-wmopen-biology1/chapter/lipids/.*

Triacylglycerol is the term used when three fatty acids are typically joined to the glycerol molecule. The amount of double bonds and chain length (i.e., total carbons in the fatty acid molecule) of the three fatty acids in triacylglycerol might vary. Below is a schematic illustration of a triacylglycerol structure, which consists of three fatty acids on a glycerol backbone (Figure 2.6).

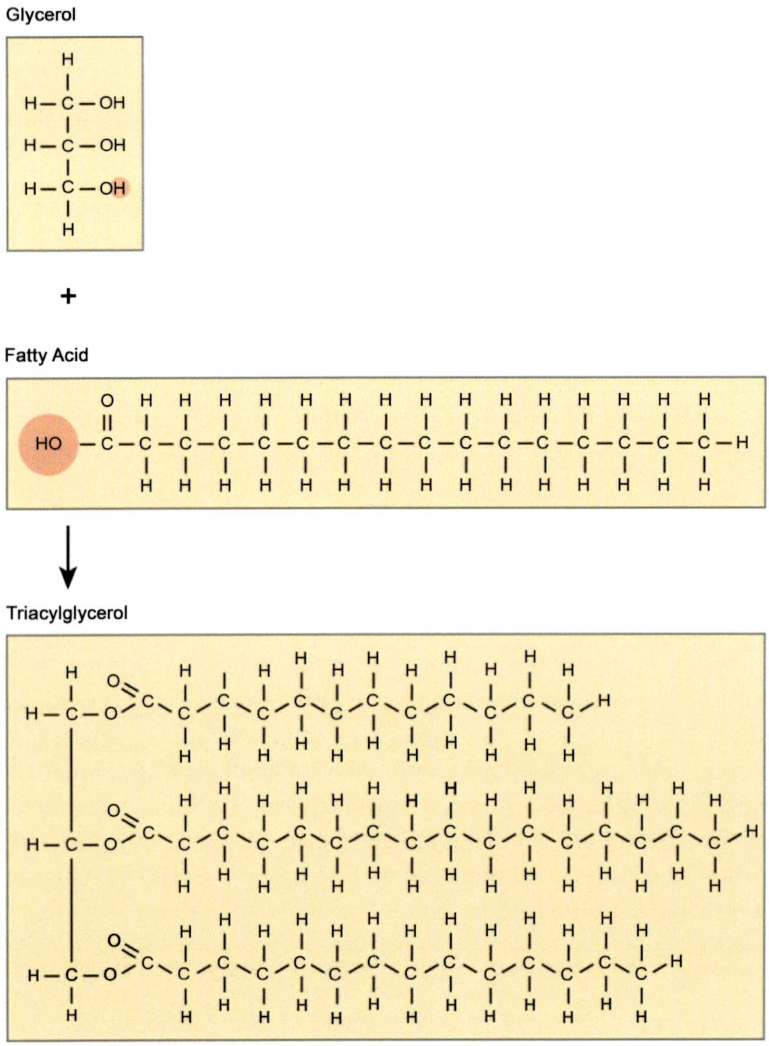


Figure 2.6. *Triglyceride structure.*

*Source: https://courses.lumenlearning.com/suny-wmopen-biology1/chapter/lipids/.*

The physical characteristics and nutritional value of fats are determined by their fatty acid content and structure. For instance, fat tends to liquefy (think salad oil) when unsaturated fats predominate in the triacylglycerol and to solidify (think fat surrounding a piece of meat) when saturated fats predominate. The presence or lack of double bonds in the hydrocarbon chain divides fatty acids into three groups.

They consist of polyunsaturated fatty acids (PUFAs), monounsaturated fatty acids, and saturated fatty acids. Saturated fatty acids, such as palmitic acid (C16:0), are "saturated" when they include hydrogen or are straight chains devoid of double bonds. Triacylglycerol often has a solid consistency when the glycerol moiety is primarily composed of saturated fats. This is because they, like tallow or beef fat, "pack" extremely firmly in the membrane because of their straight-chain nature. One or more double bonds between neighboring carbon atoms in the hydrocarbon chain are present in unsaturated fatty acids. Monounsaturated fatty acids have one double bond, whereas polyunsaturated fatty acids have more than two double bonds.

Because unsaturated lipids (like vegetable oil) have a "bend" in their structure and cannot pack as firmly as saturated fats, triglycerides with a majority of unsaturated fats tend to be watery. PUFAs, or polyunsaturated fatty acids, are fatty acids with two or more double bonds. PUFAs are more "round" in shape than saturated fats because of these additional double bonds, as opposed to the straight chain structure of a saturated fat molecule. The fat is also physically altered by these double bonds, becoming more liquid than straight-chain saturated fat. As will be discussed below, double bond location within the carbon-carbon chain plays a crucial role in nutrition and lipid metabolism in addition to their quantity.

The term omega (ω) or "n" is used by nutritionists to indicate where double bonds are located in a polyunsaturated fat (PUFA)'s carbon chain. When counting from the methyl end ($CH_3$) of the carbon chain, the omega carbon is the first carbon with a double bond. There are two PUFA classifications: omega-3 (also known as n–3 or ω-3) and omega-6 (also known as n–6 or ω-6). For example, when counting from the methyl ($CH_3$) end, omega-3 fatty acids will have the first double bond at the third carbon, and omega-6 fatty acids will have the first double bond at the sixth carbon. In certain textbooks on chemistry or biochemistry, the Greek letter Δ, usually known as "delta," also indicates the locations of the double bonds. Double bonds are indicated by the delta word with respect to the carboxyl end. But the one that dietitians most frequently use is omega, or "n" (Figure 2.7).

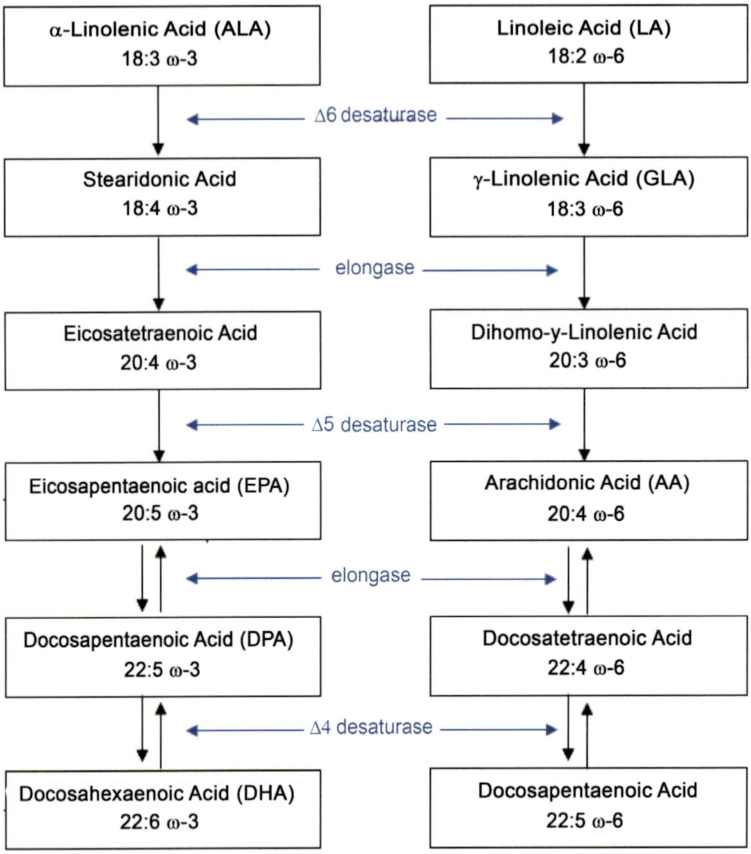


Figure 2.7. *Basic structure of an omega-3 and an omega-6 fatty acid.*

*Source: Egalini, F., Guardamagna, O., Gaggero, G., Varaldo, E., Giannone, B., Beccuti, G., Benso, A., & Broglio, F. (2023). The effects of omega 3 and omega 6 fatty acids on glucose metabolism: An updated review. Nutrients, 15(12), 2672. https://doi.org/10.3390/nu15122672.*

Two fatty acids, linoleic acid (C18:2 n–6) and α-linolenic acid (C18:3 n–3), are considered essential fatty acids and must be provided in the diet of nonruminant, or monogastric, animals like pigs. The inability to add double bonds at the third and sixth carbons from the $CH_3$ end in the n–3 and n–6 places is the cause of this essentiality. Carnivores, like cats, require arachidonic acid (C20:4 n–6), in addition to these two vital fatty acids, in their food.

## 2.2.1. Fatty Acid Nomenclature

Frequently used abbreviations for fatty acids are their shorthand notations (C18:2 n–6) or their unimportant names (linoleic acid, for example). The number of carbon atoms and double bonds are included in the abbreviated nomenclature of fatty acids. For example, C18:2 n–6 in linolenic acid represents 18 carbon atoms and 2 double bonds, the first of which is located at the carbon atom sixth from the methyl carbon. The chart below lists some of the typical fatty acids found in animal meals like chicken and pork, along with their silly names and abbreviations (Table 2.2).

**Table 2.2.** *Some Common Fatty Acids Found in Natural Fats*

| Name | Abbreviated Structural Formula | Condensed Structural Formula | Melting Point (°C) | Source |
|---|---|---|---|---|
| Lauric acid | $C_{11}H_{23}COOH$ | $CH_3(CH_2)_{10}COOH$ | 44 | Palm kernel oil |
| Myristic acid | $C_{13}H_{27}COOH$ | $CH_3(CH_2)_{12}COOH$ | 58 | Oil of nutmeg |
| Palmitic acid | $C_{15}H_{31}COOH$ | $CH_3(CH_2)_{14}COOH$ | 63 | Palm oil |
| Palmitoleic acid | $C_{15}H_{29}COOH$ | $CH_3(CH_2)_5CH{=}CH(CH_2)_7COOH$ | 0.5 | Macadamia oil |
| Stearic acid | $C_{17}H_{35}COOH$ | $CH_3(CH_2)_{16}COOH$ | 70 | Cocoa butter |
| Oleic acid | $C_{17}H_{33}COOH$ | $CH_3(CH_2)_7CH{=}CH(CH_2)_7COOH$ | 16 | Olive oil |
| Linoleic acid | $C_{17}H_{31}COOH$ | $CH_3(CH_2)_3(CH_2CH{=}CH)_2(CH_2)_7COOH$ | −5 | Canola oil |
| α-linolenic acid | $C_{17}H_{29}COOH$ | $CH_3(CH_2CH{=}CH)_3(CH_2)_7COOH$ | −11 | Flaxseed |
| Arachidonic acid | $C_{19}H_{31}COOH$ | $CH_3(CH_2)_4(CH_2CH{=}CH)_4(CH_2)_2COOH$ | −50 | Liver |

*Source: https://courses.lumenlearning.com/suny-orgbiochemistry/chapter/fatty-acids/.*

### 2.2.1.1. Cis and Trans Fatty Acids

The generation of geometric isomers by unsaturated fatty acids can be either trans or cis, contingent upon the stereo configuration of groups surrounding a double bond. Animal and plant-derived natural fatty acids are predominantly of the cis type, while bacterial-derived natural fatty acids are a combination of the cis and transforms. For instance, rumen microbes make conjugated linoleic acid (CLA), a trans fatty acid that is found in cow's milk and other ruminant foods like beef. This process is known as biohydrogenation. The two double bonds in CLA are referred to as natural trans fats since they are conjugated and do not have a methylene group separating them. Trans fats, like CLA, have drawn a lot of interest because of their many health-promoting properties, including their ability to boost immune function, prevent cancer, and increase lean body composition. When liquid vegetable oil is converted to solid fats like margarine, a process known as hydrogenation—the addition of hydrogen—occurs. This results in the production of additional trans fats. Compared to "natural" trans fats like CLA, these manufactured trans fats have differing health impacts.

#### *2.2.1.2. Cholesterol*

The most prevalent type of steroid in the human diet are lipids called steroids, which resemble rings of phenanthrene. The most well-known steroid (fat-soluble material with a steroid nucleus) is cholesterol, which also serves as a precursor to a wide range of other chemicals, including bile acids, vitamin D, sex hormones, and corticosteroid hormones. Cholesterol synthesis, which is a crucial component of animal tissues, egg yolks, and cell membranes, is partially mediated by nutrition and partially by biosynthesis from acetyl CoA. Cardiovascular diseases and the development of atherosclerotic plaque are caused by the accumulation of excess cholesterol in the arteries. Bile acid production is the process by which cholesterol is excreted. Phytosterols are a different class of sterols found in plant cells instead of cholesterol.

### 2.2.2. Digestion of Lipids

fat digestion starts in the mouth and proceeds to the stomach and small intestine, where most fat digestion occurs. The most prevalent kind of fat in blood is triglycerides. Digestive enzymes called lipases are used in lipid digestion to break down lipids.

#### *2.2.2.1. Lipid Digestion in Monogastric Animals*

Lipid molecules are broken down during the digestive process into smaller ones, which are then absorbed into the circulation. Since lipids are hydrophobic, they are insoluble in water, the aqueous medium of the digestive system. That's why dissolving fats in water is the first step in the digestion process. How? through the dispersion of lipids in tiny droplets, a process known as emulsification. Bile salt, also known as bile acid, is secreted by the gall bladder and emulsifies dietary lipids (mostly triglycerides) when they enter the small intestine. Due to its hydroxyl (OH) and carboxyl (COOH) groups, bile salt has a detergent effect, causing large fat molecules to break down into smaller lipid droplets that are encased in a bile layer. The pancreatic lipase enzyme breaks down emulsified lipids into fatty acids, monoglycerides, and glycerol. The byproducts of lipid breakdown are put together to form micelles. These are transient mixtures of monoglycerides, fatty acids, bile salt, and other fat-soluble materials like cholesterol and vitamins. Because the micelles dissolve in water, they can carry the byproducts of lipid digestion to the surface of the small intestine, where they can be absorbed. The bile salt returns to the colon to continue the emulsification processes (bile salt recycling) when the micelle disintegrates at the site of absorption. Passive diffusion allows the components to enter the small intestine and be absorbed. In summary, bile salt and the capacity to form micelles are critical for the digestion of lipids, and their absence can impact digestibility. For instance, unsaturated fatty acids form micelles more effectively than saturated fatty acids. Therefore, animal meals contain a combination of unsaturated and saturated fatty acids. The monoglycerides and fatty acids are re-esterified once they are within the intestinal cell (or enterocyte), where they are combined with free and esterified cholesterol, lipoproteins, and phospholipids to form chylomicrons. The lymphatic system receives the secretion of chylomicrons.

### *2.2.2.2. Lipid Digestion in Ruminant Animals*

Low (less than 5%) fat is found in the diets of ruminant animals, and it can be found in a variety of foods, including cereal grains, grass, leaves, and oil seeds. Oil seed or grain lipids are mostly composed of triglycerides, while leaf or grass lipids are primarily composed of galactolipids, phospholipids, waxes, pigments, and essential oils. Neither pancreatic lipase enzyme nor emulsifying agent exists in the rumen. Microbial lipases are instead produced by rumen bacteria. The hydrolysis of ester bonds in triglycerides, phospholipids, and glycolipids occurs first when dietary lipids enter the rumen. Microbial lipases hydrolyze dietary lipids, releasing glycerol and free fatty acids from the lipid backbone in the process. The rumen bacteria easily break down glycerol to produce propionic acid. The proportion of propionic acid, one of the volatile fatty acids, or VFAs, and the propionate:acetate ratio rise when ruminants are fed additional fat. The following step requires hydrolysis. Unsaturated fatty acid biohydrogenation is the second significant change that dietary lipids can go through in the rumen. Microbes change double-bonded fatty acids to create fatty acids that are more stable. Conjugated fatty acids, such as conjugated linoleic acid, or CLA, are derived from fatty acids like linoleic acid in which the double bonds are not broken up by methylene ($CH_2$) groups. The fatty acids are changed into more stable "trans" fats by moving the double bonds. During this process, several branched-chain and odd-numbered fatty acids (such C19:0) are also produced. For instance, during this conversion stage, linoleic acid (C18:2 n–6), where the double bonds are in the cis position (cis9-cis12), is transformed into several isomers of CLAs. The small intestine of ruminants' digests fats very similarly to the intestines of monogastric animals. Pancreatic juices and bile are the two secretions that are essential to this procedure. The lipids can form micelles for absorption thanks to these secretions. Pancreatic juice, enzymes, and bile salts are all obtained from bile. These substances enable the production of micelles by desorbing the fatty acids from bacteria and feed particles. When micelles develop, they help move lipids that are insoluble in water between the intestinal epithelial cells of the jejunum and the absorption site of fatty acids. Fatty acids are re-esterified into triglycerides within the intestinal epithelial cells, and they are then packed into chylomicrons for lymphatic transport to the blood.

## 2.2.3. Lipids Utilization

### *2.2.3.1. Transportation*

Blood lipids are made up of lipids from storage sites such as adipose tissue and liver, as well as chylomicrons that are created within the intestinal mucosal cells during absorption. Because blood lipids are hydrophobic, they are carried as lipoproteins. Lipoproteins are composed of an exterior layer of proteins (apolipoprotein) and phospholipids and cholesterol surrounding an inner core of hydrophobic lipids. A lipid plus a protein is called a lipoprotein (compound lipid). Lipids are less dense than water, so as the ratio of lipid to protein rises, lipoprotein density falls. Lipoproteins play a crucial part in the metabolism and transport of lipids, including cholesterol. Lipoproteins are categorized according to their composition and density. Chylomicrons, very-low-density lipoproteins (VLDL), low-density lipoproteins (LDL), and high-density lipoproteins (HDL) are the four main lipoproteins found in blood. The small intestine produces chylomicrons from dietary fat, whereas the liver and small intestine produce VLDL, LDL, and HDL. After entering the liver, chylomicrons are contained in VLDL. Triacylglycerol (TAG) is transported from the liver to extrahepatic tissues by VLDL. HDL is in charge of "reverse transport," or removing cholesterol (also known as good cholesterol) from tissues, while LDL is responsible for delivering cholesterol (also known as bad cholesterol) to tissues. "Bad" and "good" describe different aspects of transportation. The term "good" cholesterol, for instance, describes cholesterol carried by HDL, which is intended to be excreted from the body through the liver for the production of bile rather than being deposited into blood vessels as LDL does.

### *2.2.3.2. Metabolism*

Because it is involved in the synthesis and catabolism of lipoproteins, the liver plays a crucial role in lipid transport and metabolism. Because acetyl CoA plays a crucial role in both lipid and carbohydrate metabolism, mechanisms of fatty acid breakdown and biosynthesis are closely linked to pathways of carbohydrate metabolism. Blood is used to carry the fatty acids that are liberated during TAG hydrolysis. The process of oxidation takes place in the mitochondria of several tissues, including skeletal muscle, heart, liver, and adipose tissue. This conversion results in the production of acetyl CoA, which is subsequently broken down in the citric acid cycle. Fatty acids are activated (they react with CoA) to create fatty acyl CoA before oxidation starting (ATP is needed for this step). Carnitine acts as a transporter to transfer activated fatty acids across the mitochondrial membrane. In β-oxidation, fatty acid CoA is broken down into two carbons at a time, beginning at the carboxyl end. High yields of ATP are created when hydrogen released from fatty acids is absorbed up the chain by hydrogen acceptors (such as FAD and NAD+). When released, acetyl CoA can be used to resynthesize fatty acids, produce ketone bodies, or enter the tricarboxylic acid cycle (TCA) for the synthesis of ATP. The β-oxidation pathway is used to oxidize 16 and 18-carbon fatty acids. Within the peroxisomes, additional longer-chain fatty acids undergo oxidation. Unsaturated fatty acids undergo β-oxidation up until they encounter their first double bond. Acetyl CoA is finally produced by the hydrolysis of the double bonds by isomerase once they are isomerized from cis to trans.

The process of creating lipids to store chemical energy is known as lipogenesis. Adipocytes, or fat cells, are found all over the body and are thought to be a long-term energy store. Lipogenesis is the process by which lipids are stored in the body as a "savings account," adipocyte absorption, and fatty acid production (cytosol of hepatocytes and adipocytes). The primary cue for fat storage (feasting) is glucose. The body absorbs excess energy (ATP) produced by the TCA cycle and glycolysis and stores it as fat. When ATP levels rise over what the cells require, ATP starts to build up and the enzyme acetyl CoA carboxylase is activated. To promote acetyl CoA carboxylase activity, higher insulin concentrations are also necessary. Similar to β oxidation, 2 C acetyl CoA initiates fatty acid production. Certain amino acids, lipids, and carbs can all provide acetyl CoA. The cytoplasm is where fatty acid production takes place (endoplasmic reticulum). By connecting the carboxyl end of one fragment to the methyl tail of another, the fatty acid chain is formed in two C units (produced from acetyl CoA), with palmitic acid (C16:0) being the final product. In short, one mole of acetyl CoA is combined with one ATP by acetyl CoA carboxylase to generate malonyl CoA. In order to create 3 C malonyl CoA, carbon dioxide must be added to 2 C acetyl CoA to start the fatty acid production process. This step requires the enzyme acetyl CoA carboxylase. A 5 C intermediate molecule is created when malonyl CoA combines with acetyl CoA. This product is then decarboxylated to create a 4 C butyryl CoA. Subsequently, this amalgamates with an additional malonyl CoA and undergoes decarboxylation to generate 6 C caproyl CoA. The process is then repeated, adding malonyl CoA and exchanging $CO_2$ each time, yielding the ultimate result of 16 C palmitic acid (Figure 2.8).

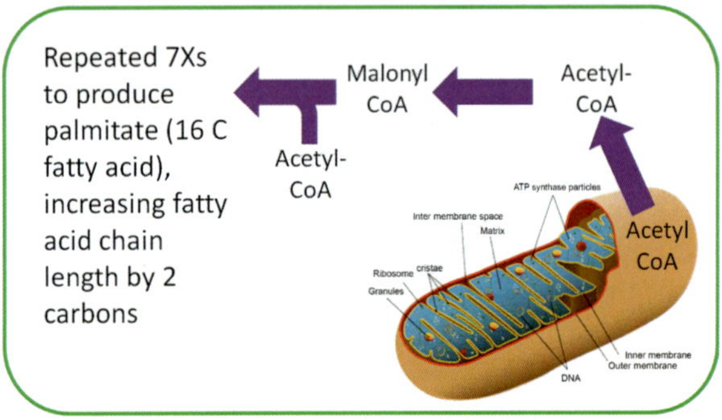


Figure 2.8. *Fatty acid biosynthesis.*

*Source: https://courses.lumenlearning.com/suny-nutrition/chapter/6-33-de-novo-lipogenesis-fatty-acid-synthesis/.*

Because each acetyl CoA and subsequent malonyl CoA only have two carbons, most fatty acids are even numbered. With every additional carbon added by a malonyl CoA, the ensuing fatty acid is nearly always linked evenly. Animal tissues contain very small amounts of odd-numbered fatty acids (C17:0, C19:0, etc.). The enzyme in these processes is called fatty acid synthase. Elongases have the ability to extend palmitic acid (16 C) into 18 C stearic acid. The addition of double bonds by the desaturase enzyme results in the further synthesis of unsaturated fatty acid (oleic acid), C18:1. Double bonds cannot be inserted by mammals beyond the Δ-9 position. Thus, the necessary fatty acids linoleic (18:2 Δ-9,12) and linolenic (18:3 Δ-9,12,15) must be obtained by diet. Mammals may, however, further desaturate and extend the 18 C essential fatty acids after eating them in order to create longer chains of 20 and 22 C fatty acids.

### *2.2.3.3. Storage*

TAG is used to store lipids. Together with glyceraldehyde-3-phosphate, dihydroxyacetone phosphate is generated during the preparation stage of glycolysis and forms the glycerol backbone of TAG. Since glycerol has three carbons, a fatty acid can form a bond with any one of the carbons. In ruminant adipose tissue or the cytoplasm of hepatocytes, glycerol, and fatty acids mostly link (or the mammary gland in breastfeeding animals). After forming, TAGs are deposited in the bloodstream for transportation after adhering to lipoproteins. Lipoprotein lipase recognizes the lipoprotein complexes, and the TAGs are extracted and deposited as fat (e.g., back fat in pigs, fat pads in hens). Hormone-sensitive lipoprotein lipase releases free fatty acids and glycerol when it mobilizes lipids from storage sources for energy generation. Starvation (fasting), stress, or increases in energy consumption (illness, for example) all lead to mobilization. In these circumstances, hormones like glucagon and adrenaline rise, insulin falls, and lipases (hormone-sensitive enzymes) are stimulated. Fat mobilization peaks in dairy cattle and ewes under specific situations, such as negative energy balance, and can outpace the rate at which acetyl CoA enters the TCA cycle. Ketogenesis, or the synthesis of ketone bodies in the liver, takes place under such circumstances. Acetone, β-hydroxybutyric acid, and acetoacetic acid are the three ketone bodies. When too many ketone bodies are created, an illness known as ketosis results.

### *2.2.3.4. Ruminant Animals*

Instead of using glucose, ruminant animals immediately get acetyl CoA from ingested acetate (volatile fatty acid). Because citrate is permeable, it can pass through the mitochondrial membrane and be transformed into α-ketoglutarate when NADPH, nicotinamide adenine dinucleotide phosphate, is produced. While fatty acid synthesis from acetate is being carried out by NADPH, the α ketoglutarate returns to the mitochondria. Ruminants use it as a way to store glucose and glucogenic precursors.

## 2.3. PROTEINS

The Greek term "proteios," which means first or most important, is where the word "protein" originates, and it was coined by Dutch chemist G. J. Mulder. Proteins are organic substances composed of several amino acid building blocks, or basic units, bonded together by peptide bonds. A tripeptide has two peptide bonds and three amino acids, while a dipeptide has one peptide bond and two amino acids. A polypeptide is a peptide that has more than 10 amino acids. Large polypeptides are basically what proteins are. The order of individual amino acids in a polypeptide chain determines a protein's shape initially. This is often referred to as the protein's main structure.

### 2.3.1. Functions of Protein

The main structural elements of animal tissues, such as skin, muscles, wool, feathers, tendons, and eggs, are proteins, which are essential for life. Furthermore, proteins have roles in transportation (lipoproteins), immunological responses (immunoglobulins), metabolic reactions (enzymes), and other regulatory functions (hormones). When required, proteins can also supply energy. Proteins catalyze many of the structures seen in animal tissue, such as muscle, as well as metabolic processes, such as enzymes and hormones. Thus, the maintenance of the life process depends on protein synthesis. Enough protein and amino acids in the diet are necessary to support growth, health, and productivity in animals raised for food. In ruminant animals, the intestinal microbiota can synthesize proteins from nonprotein sources. The amount of protein needed varies depending on the stage of life; in young animals, it is higher during periods of rapid growth and during pregnancy and lactation. Proteins are composed of hydrogen, oxygen, and carbon, just like other macronutrients. Furthermore, proteins include sulfur and nitrogen (in some amino acids). Regarding their ability to be absorbed, metabolized, and eliminated by the animal, proteins are highly distinctive among other foods in animal nutrition because of their nitrogen content.

### 2.3.2. Proteins Classification

Proteins can be categorized by shape, prosthetic group composition, solubility in water, salt, acid, base, or alcohol, or all of the above.

#### *2.3.2.1. Classification Based on Solubility and Prosthetic Group*

1. **Globular Proteins:** One of the common protein types in biochemistry, globular proteins, also known as sphero-proteins, are spherical ("globe-like") proteins (the others being fibrous, disordered, and membrane proteins). Unlike fibrous or membrane proteins, globular proteins form colloids in water and are slightly water soluble. Since there are numerous distinct architectures that can fold into a roughly spherical shape, globular proteins can be classified into multiple fold classes. More precisely, the term "globin" can refer to proteins that include the globin fold. Water-soluble or diluted in alcohol, bases, or acids:
   - Water-soluble albumin is found in egg whites as albumen and serves a variety of purposes in blood circulation, including transporting lipids.
   - Globulin, soluble in diluted neutral solutions, plays a defensive

role in the body as a component of the immune system (e.g., immunoglobulins).

2. **Fibrous Proteins:** Together with globular and membrane proteins, fibrous proteins, also known as scleroproteins, are one of the three primary categories of protein structure in molecular biology. Polypeptide chains that are stretched or fibrous combine to produce filamentous and sheet-like structures that make up fibrous proteins. These proteins have a lower water solubility than globular proteins, which helps to differentiate them. By producing bone matrices, tendons, muscle fiber, and connective tissue, these proteins fulfill structural and defensive functions. The superfamilies of fibrous proteins include keratin, collagen, elastin, and fibrin. The most prevalent of these proteins in vertebrate connective tissue, such as tendon, cartilage, and bone, is collagen. They are resistant to digestive enzymes and insoluble in water. Although these proteins don't offer much nutritional value, they are crucial for many metabolic, structural, and biochemical processes. Feather meal, for instance, has a high protein content (keratin) but a relatively low digestion, making it a limited feed item for animals. Feather meal's polypeptide chain's amino acids combine to form disulfide bonds (-S-S-), which bend the chain into a particular coiled shape like a helix or sheet. We refer to this as a secondary structure. The hard physical characteristics and poor digestion of hooves and horns are explained by these connections.
   - Keratins (e.g., wool, hair, feather, hooves, horn);
   - Collagen (can be converted to gelatin when heated; present in bone, teeth, tendons, and soft connective tissue).

3. **Conjugated Proteins:** These are proteins that interact with other (non-polypeptide) chemical groups through weak contacts or covalent bonds. Simple proteins, on the other hand, are composed entirely of amino acids without any additional chemical groups. Conjugated proteins, upon hydrolysis, produce a different chemical component in addition to amino acids. The non-amino portion of a conjugated protein is referred to as its prosthetic group. Most prosthetic groups are derived from vitamins. The prosthetic groups of conjugated proteins are categorized based on their chemical makeup and contain structurally non-protein substances. Here are a few examples:
   - Lipoproteins (lipid-carrying protein);
   - Hemoprotein (proteins with heme units);

- Glycoproteins (proteins with sugar);
- Nucleoproteins (proteins bound to nucleic acid).

Denaturation of the proteins results from heat treatment's disruption of secondary structure (e.g., egg white coagulation after cooking). Proteins are present in several antinutritional components of feed, such as the trypsin inhibitor found in soybean meal. The trypsin inhibitor in soybean meal is denatured by heat processing, which also improves digestibility.

## 2.3.3. Amino Acids

The building components of proteins are amino acids. It is known that nature contains more than 300 different types of amino acids. Of these, roughly 20 amino acids are crucial components of animal proteins and are linked to blood, hormones, muscles, connective tissues, skin, feathers, horns, and enzymes. The term "essential amino acids" refers to a group of around 10 amino acids that are required for proper metabolic processes in animal tissues, but which cannot be synthesized by animal tissues. Certain additional amino acids, including citrulline and ornithine, are involved in cellular metabolism but are not found in animal tissues. By definition, every amino acid has a minimum of one carboxyl group (–COOH) on the C atom next to the carboxyl group and one amino group ($-NH_2$). Proline, an amino acid, is an exception to this rule because it doesn't have a free amino group. The most basic amino acid, glycine, is displayed below to illustrate the typical structure of an amino acid. For each amino acid, the R group (shown by the red circle) differs. The remaining portion of the molecule or any other group joined to the C atom is known as the R group. It is a H group in the instance of glycine. The carboxyl (COOH) group gives amino acids their acidic qualities, whereas the amino group ($NH_2$) gives them their basic properties. Alpha (α) amino acids, carboxylic acids with an amino group on the α-carbon (or the first carbon connected to a functional group), are essential amino acids for animal nutrition (Figure 2.9). Figure 2.10 shows the essential amino acids for animal nutrition along with their classification and rank of importance.

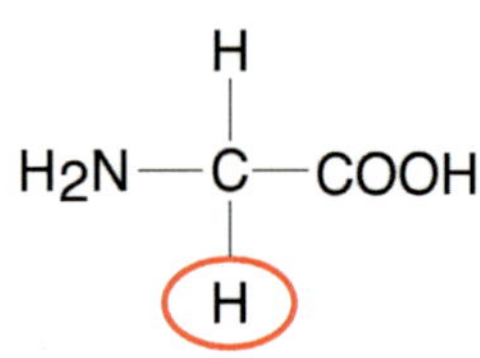


Figure 2.9. *General structure of amino acid.*

*Source: Nasef et al. (2020).*

There are two isomeric forms of amino acids: D- and L-isomers. The arrangement of groups around the asymmetric α-carbon varies in D and L amino acids. With the exception of methionine, where an animal can use both D- and L-amino acids, only L-amino acids are utilized in the production of proteins. Animal feeds frequently include DL-methionine as an amino acid supplement. Except for glycine, every amino acid has an asymmetric α-carbon, which has four distinct chemical groups connected to it. Isomers of compounds containing asymmetric carbons are possible (Figure 2.10).

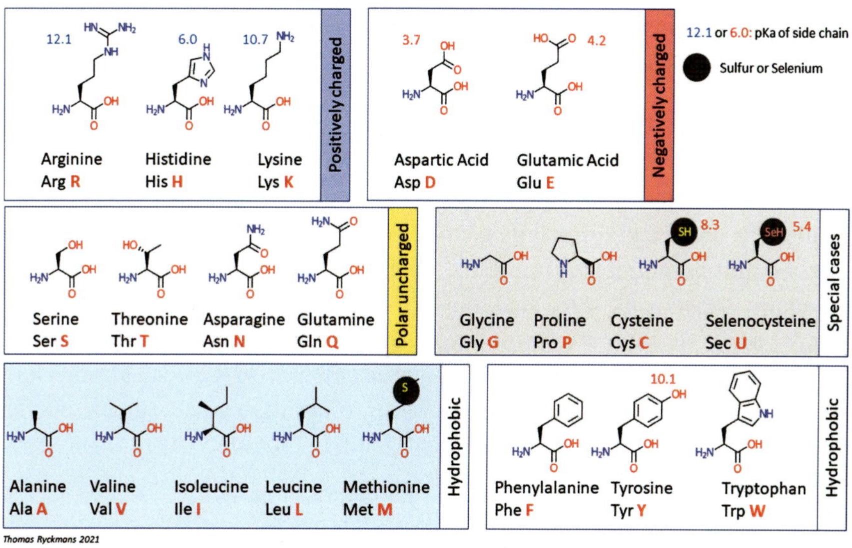


Figure 2.10. *Classification of amino acids.*

*Source: By Thomas.ryckmans68 – Own work, CC BY-SA 4.0, https://commons.wikimedia.org/w/index.php?curid=117084381.*

## *2.3.3.1. Essential Amino Acids*

A portion of amino acids can be sufficiently synthesized by animal bodies. Nevertheless, certain amino acids are either not synthesized by animals or not in sufficient amounts to meet their bodies' needs. These amino acids are referred to as essential (indispensable) amino acids because they must be obtained from the diet of animals with a single stomach. Around 10 essential amino acids are needed by people, dogs, and pigs, whereas 11 essential amino acids are needed by chickens and cats. The following is a list of the necessary amino acids required by monogastric animals.

- Arginine (Arg);
- Histidine (His);
- Lysine (Lys);
- Isoleucine (Ile);
- Leucine (Leu);
- Methionine (Met);
- Phenylalanine (Phe);
- Threonine (Thr);
- Tryptophan (Try);
- Valine (Val).

Cats and chickens require the following additional amino acids in addition to the 10 essential amino acids.

Chickens require glycine (Gly), while cats require taurine (Tau). Remember that other non-essential amino acids, such as carbohydrates and nonprotein nitrogenous substances, are also physiologically significant for the body's metabolic processes and are formed from different precursors that are obtained from the diet. Animals have different requirements for critical amino acids. Horses, for instance, require critical amino acids, but since rumen bacteria synthesize these amino acids, ruminant animals (cattle, sheep, and goats) typically do not.

## 2.3.4. Protein Utilization

### *2.3.4.1. Digestion*

The process of breaking down ingested feed into simple compounds that the digestive tract can absorb is called digestion. Proteins must first be denaturized in order to reveal their peptide bonds, after which they must hydrolyze and release free amino acids. Endopeptidase and exopeptidase are two types of enzymes that break down proteins. Peptide bonds inside the main structure are broken into smaller fragments by endopeptidases. Amino acids are separated from the protein molecule's terminal end by exopeptidases. Aminopeptidase acts on the terminal amino acid with a free amino group, while carboxypeptidases remove an amino acid from the end with a free carboxyl group. Digestion of proteins starts in the stomach. The hormone called gastrin starts the process of breaking down proteins in the stomach. The main cells of the gastric mucosa secrete pepsinogen when food is present in the stomach. Pepsinogen is converted into the active form of pepsin by the stomach mucosa's parietal cells using HCl. An endopeptidase is pepsin. In young animals, the stomach secretes milk coagulating rennin, which helps form clots and facilitates passage into the small intestine. The small intestine, which is crucial for the digestion of proteins, is where the next stage of digestion takes place. Trypsinogen, chymotrypsinogen, and procarboxypeptidase are three of the inactive forms of the enzymatic secretions from the pancreas that are stimulated by the hormone secretin in the duodenum. Additionally produced at the duodenum, enterokinase changes trypsinogen into trypsin, which in turn changes pro-carboxy-peptidase and chymotrypsinogen into their active forms, chymotrypsin, and carboxypeptidase. Other enzymes, such as dipeptidases and aminopeptidases from mucosal membranes, complete digestion. This procedure aims to reduce polypeptides to single-chain, free amino acids. The small intestine's villi aid in the absorption of substances into the bloodstream, just as they do with fats and carbs. Sodium is a type of cotransported chemical that is used by normal free proteins during their energy-intensive, active transport process. Whole proteins move materials directly, using no energy at all. The primary type of amino acids that are absorbed into the circulatory system is free form. A small amount of tri-, di-, and oligopeptides are also absorbed, though. The amino acid transfer involves specific carrier proteins depending on the kind of amino acid (neutral, basic, acidic, big, tiny, etc.). Absorption of amino acid forms is preferred for the naturally occurring L-forms over the D-forms. Certain amino acids might compete with others for transport and carrier proteins. For instance, high leucine concentrations increase the need for isoleucine, while arginine inhibits lysine transport. A few neutral amino acids prevent the transfer of basic amino acids.

### *2.3.4.2. What Happens to Amino Acids?*

The production of tissue proteins, enzymes, and hormones, and the deamination or transamination of carbon skeletons can all be accomplished with the help of absorbed amino acids. Ammonia and other polyamines are produced in the hindgut as a result of microbial fermentation of undigested proteins.

There are two stages of protein digestion in ruminant animals: (1) reticulorumen digestion (degradation) and (2) abomasum and small intestine digestion. As a result, dietary proteins in ruminant animals are divided into two categories: proteins that degrade in the rumen and proteins that do not. The primary objective of protein supplementation, similar to that of monogastric animals, is to supply the animal with amino acids. But in ruminants, proteins provide the rumen microorganisms with nitrogen, which allows them to produce their own microbial protein from scratch. Microbes can employ nitrogenous molecules that are not proteins, like urea, for the synthesis of microbial proteins without giving a damn about where the nitrogen comes from. Microbial urease in the rumen completely breaks down urea (but greater concentrations may be hazardous). Both bacteria and protozoa, which generate proteolytic enzymes, can break down the protein that enters the rumen. Proteases and peptidases, which cleave peptide bonds in polypeptides to release free amino acids from proteins, are produced by rumen bacteria. Rumen breakdown can be influenced by many variables, including protein solubility and structure. Through a process known as deamination, these amino acids broken down by the rumen liberate $NH_3$ and the C skeleton. The main source of protein for the host ruminant animals is the microbial protein that is produced by rumen bacteria in addition to volatile fatty acids (from carbs). For animals that produce in large quantities, microbial protein is insufficient for sustenance and survival. Blood urea nitrogen (BUN) is the product of the conversion of ammonia from rumen to urea and its secretion into the blood. Saliva or the rumen wall can filter and recycle urea into the rumen. The level of BUN in ruminants is an indicator of how well they use protein. Proteins with poor rumen degradation rates, such as those found in maize, are referred to as "escaped," "bypassed," or "undegradable" (rumen undegradable protein, RUP). RUP reaches the ruminant animal's small intestine and abomasum for absorption and digesting. Proteins originating from microbial sources or RUP may make their way into the small intestine. Microbial proteins and RUP supply the host animal's amino acid requirements. The essential amino acids are needed by both monogastric and ruminant animals and must be continuously supplied by food because the body is unable to store them. Table 2.3 illustrates some of the similarities and differences between ruminant and monogastric animals with regard to the breakdown or digestion of proteins.

Table 2.3. *Similarities and Differences in Monogastric and Ruminant Animals in Protein Digestion*

| **Monogastric** | **Differences (Ruminants)** |
|---|---|
| Amino acid profile at small intestine reflects the diet. | Amino acid profile at the small intestine is different from diet. |
| No upgrading of low-quality dietary protein. | Up-grade low quality dietary protein. |
| Protein quality not downgraded. | Down-grade high quality dietary protein. |
| Cannot use non-protein nitrogen. | Able to use non-protein nitrogen (e.g., urea). |
| Constant supply of amino acids is required. | Constant supply of amino acids is required. |

*Source: Cherian (2020).*

Corn has the largest bypass potential among cereal grains. It should be mentioned, though, that maize lacks many vital amino acids, like methionine and lysine. Fish and meat meals, two forms of animal protein, have a significant bypass potential. Bypass potential is increased by heat treatment and drying of forages. Because heat is generated during feed processing techniques like pelleting, steam rolling, or flaking, the feed protein is often denatured, "protecting" it from lysis in the rumen. Commercially available rumen protected protein sources dissolve in the abomasum and stay intact in the rumen after being treated with formaldehyde.

### *2.3.4.3. Metabolism*

Every protein in the body is constantly changing, and the ratio of synthesis to breakdown determines how big the pool of amino acids is. The production of non-essential amino acids, tissue proteins, enzyme or hormone synthesis, deamination, and transamination are examples of anabolic uses for absorbed proteins. The primary location of amino acid metabolism is the liver. Through a process known as transamination, the liver synthesizes nonessential amino acids and possesses enzymes like transaminases. An amino group from one amino acid is transferred to an organic acid in this process, creating a new amino acid. Pyridoxine, or vitamin B6, is required for transaminase function. Additionally, transamination establishes a connection between the metabolism of proteins and carbohydrates by allowing some amino acids to employ their C skeleton to synthesize glucose. An amino acids amino groups are removed during deamination to produce ammonia. The animal's body needs to expunge nitrogen, which requires this procedure. C skeletons are what remain after deamination or transamination and are utilized to produce glucose, ketone bodies, or energy. All amino acids can exploit the C skeleton for the production of glucose, with the exception of leucine and lysine. As strictly ketogenic amino acids, leucine, and lysine can produce ketone bodies and acetyl CoA, which can be used as an energy source. Acetyl CoA cannot produce glucose because it loses both of its carbons during the tricarboxylic acid cycle (TCA). Tryptophan, phenylalanine, tyrosine, and isoleucine are among the amino acids that are both glucogenic and ketogenic.

Because it is hazardous to the central nervous system, the ammonia released during the breakdown of amino acids must be eliminated or detoxified. While birds excrete ammonia as uric acid (a white component in the excreta), most mammals detoxify it and excrete it as urea in the urine. The liver and kidney are the two tissues that the urea cycle passes through to detoxify ammonia and produce urea. Ornithine and citrulline are two nonprotein amino acids—amino acids not utilized in the creation of proteins—that are important in the urea cycle. The liver's mitochondria condense ammonium ions with bicarbonate ions to generate carbamoyl phosphate, which is the first step in the urea cycle (Figure 2.11).

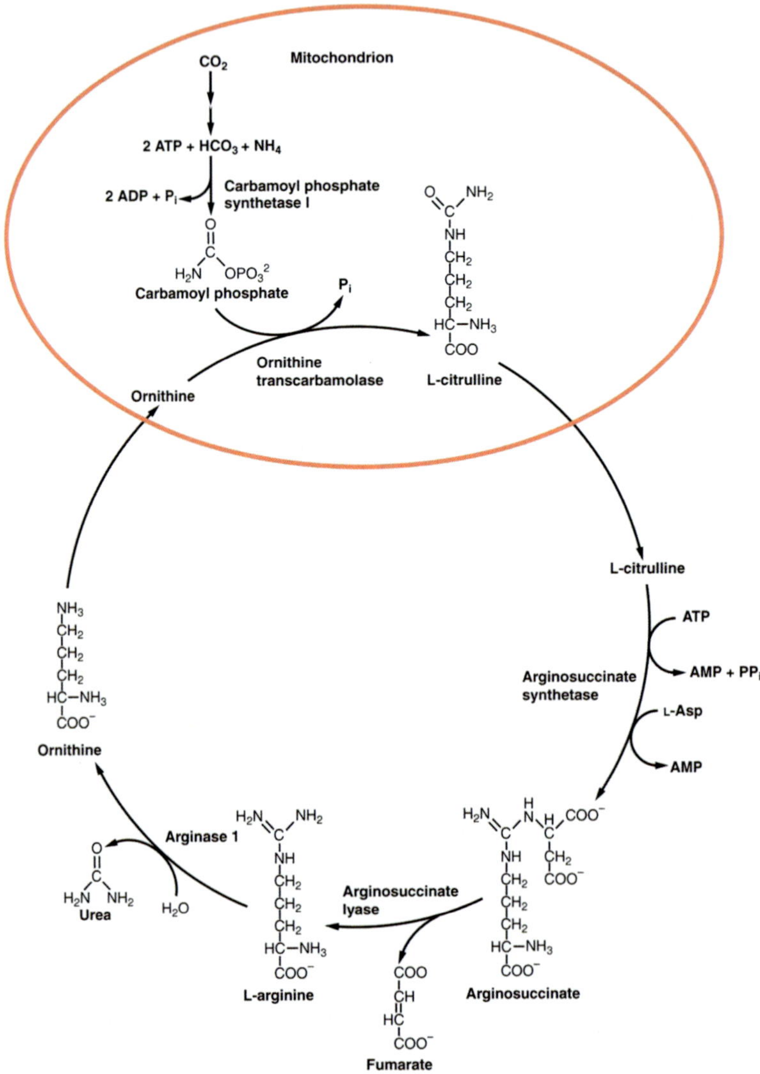


Figure 2.11. *Urea cycle.*

*Source: https://courses.lumenlearning.com/suny-ap2/chapter/protein-metabolism/.*

Citrulline is created when ornithine combines with a substance known as carbamoyl phosphate. Because citrulline is highly soluble, it enters the cytosol and combines with aspartate to create argininosuccinate, which requires two ATPs to metabolize. Argininosuccinate is broken down by the enzyme argininosuccinate lyase into arginine and fumarate, which then enters the TCA cycle. To restart the cycle, arginine is lysed into ornithine, which then splits off urea to produce ornithine. As arginine isn't needed for protein formation, it can be considered a nonessential amino acid.

Because they are unable to synthesize carbamoyl phosphate, poultry is unable to produce urea. Instead, the synthesis of uric acid is carried out using glutamic acid, glycine, and methionine. Therefore, for optimal production, chickens need a diet rich in methionine, arginine, and glycine. Since the urea cycle depends on ATP for urea synthesis, feeding animals excessive or low-quality protein requires a lot of energy and might have negative environmental effects (such as groundwater contamination and air ammonia). Ammonia ($NH_3$) in the rumen of ruminant animals is transported via the portal vein to the liver, where it undergoes the urea cycle to be converted into urea and released into the bloodstream. From there, it is either excreted in the urine or returned to the digestive system to serve as a nitrogen source for rumen microorganisms. Every bodily tissue is involved in the creation of proteins. The genetic code is ultimately responsible for protein production. The specifics of protein synthesis are covered in other biochemistry literature and are outside the purview of this chapter. Nucleotides are the building blocks of nucleic acids, which include ribonucleic acid (RNA) and deoxyribonucleic acid (DNA), which are molecules that carry genetic information. DNA is the building block for the creation of proteins and holds the animal's genetic information.

The synthesis of RNA—a transcription template—is regulated by DNA. Messenger RNA, transfer RNA, and ribosomal RNA are the three kinds of RNA. Protein synthesis involves all three of them. The portion of ribosomes called ribosomal RNA has three base codons that code for amino acids (the building blocks of proteins). Certain amino acids are taken up by transfer RNA from the cytoplasm and sent to the ribosomes, the protein-making machinery. Transfer RNA (tRNA) functions as an adaptor when proteins are being synthesized. Every tRNA has a unique amino acid carrying it. The sequence of amino acids (translation) in the protein that is generated is determined by messenger RNA. The DNA template is used to make messenger RNA (mRNA) and tRNA. Different mRNAs regulate the production of different proteins. The inability of an empty space to occur during the production of the peptide chain restricts the formation of peptide chains and the synthesis of proteins. It takes all 20 amino acids to synthesize proteins. For instance, a diet deficient in necessary amino acids can prevent the creation of peptide chains and protein synthesis, which can impact an animal's ability to acquire weight and function. Protein synthesis and breakdown occur continuously and simultaneously during the dynamic process known as protein turnover. The balance between synthesis and degenerative processes determines the net rate of protein growth or loss. The reason for the body's constant need for proteins is the cycling of proteins and the loss of proteins, primarily in feces. An animal still needs protein even when it is not growing. Age, physiological (e.g., pregnancy, breastfeeding) and pathological status, as well as the quality of the protein consumed, all influence how much protein is required in the diet.

## 2.3.5. Quality of Proteins

For effective, affordable, and sustainable feeding, the estimation of the nutritional needs of cattle and the assessment of the degree to which various feedstuff sources offer necessary nutrients are required. This context involves a detailed investigation into the calculation of protein requirements and the degree to which essential and nonessential amino acids are supplied by various protein sources. The availability of amino acids that a protein delivers is referred to as its quality, while digestibility takes into account the optimum way to use the protein. The quality of different kinds of protein can be ranked using a variety of techniques; this chapter outlines some of the in vitro laboratory studies and in vivo animal bioassays that are frequently employed in animal nutrition. The quality of protein can be impacted by several things.

- Amino acid profile;
- Content and balance of essential and nonessential amino acids;
- Content of limiting amino acids;
- Protein digestibility and bioavailability.

The requirements of animals are for amino acids rather than proteins. The primary purpose of dietary protein is to provide the necessary amounts of the amino acids. Thus, the amino acid profile of the diet, the amount of limiting as well as essential amino acids, the digestibility of amino acids after digestion, and their physiological utilization all influence the quality of feed protein. For nonruminants, these features are especially crucial. For ruminant animals, solubility and digestibility are also crucial. The percentage of protein that is consumed and absorbed by the animal—that is, not expelled in feces—is known as the digestibility of protein. Therefore, by measuring dietary protein intake and fecal output, one can compute digestibility, as indicated by the following equation:

Amino acid digestibility (%) = Amino acid consumed − Amino acid excreted × 100 / Amino acid consumed

The preferred method for determining the availability of amino acids is to use digestibility assays. The majority of commonly used and published values on the digestibility of protein and amino acids in monogastric animals are derived from fecal or excreta analysis due to its ease of use and high sample size, which allows for the use of many animals without requiring animal sacrifice. Yet, the hindgut's altering and fluctuating impacts on fecal-based digestibility assessments (microbial utilization of dietary protein and the contribution of microbial and endogenous proteins to amino acid excretion in feces contributing to fecal amino acid output) are a problem. There are two categories for digestibility: ileal and fecal. Ileal digestibility assays are performed by digesta collection at the ileal junction, either with cannulated animals or by slaughtering small animals like chickens. Fecal digestibility assays are conducted through fecal collection and examination. Bioavailability and digestibility are frequently used interchangeably. This is untrue, though. An amino acid that is in a state that is suited for digestion, absorption, and utilization is said to be bioavailable. Only animal growth bioassays using live animals can be used to determine bioavailability. The rate at which amino acids are absorbed and become available for protein synthesis is known as their bioavailability. Digestion and bioavailability can be impacted by a number of dietary and animal-related factors. Protein digestibility may be impacted, for instance, by the protein's source (animal versus plant) and the feed processing techniques used. The essential and nonessential amino acids found in animal proteins are more evenly distributed. Protein digestibility and amino acid availability can be impacted by antinutritional agents found in plant proteins, such as trypsin inhibitors, found in soybean meal. Feed processing techniques like roasting, heating, and extrusion can damage soybean meal's trypsin inhibitor and other antinutritional components. However, a complicated combination with sugars causes warming (also known as browning or the Maillard reaction), which renders amino acids inaccessible. Amino acid imbalance can also be caused by other interactions, such as antagonistic associations, where excesses of one amino acid interfere with the metabolism of another, rendering it inaccessible (e.g., arginine and lysine).

### *2.3.5.1. Methods for Protein Quality Assessment*

It is well known that the nutritional value of proteins varies widely. This may be blatantly shown by a variety of techniques, including comparing the growth rates, nitrogen retention, and other physiological performance measurements of animals or humans fed diets with about similar proportions of various proteins. It's also evident that these variations are typically caused by the amino acid makeup of the proteins, as adding important amino acids to proteins frequently results in a significant increase in their nutritional value. It has long been believed that some of these nutritional quality indicators were accurate and suitable enough to calculate the required amount of protein when different proteins were consumed if the amount of a specific protein was known. Therefore, it has been generally advised to estimate protein requirements using the highest-utilized dietary protein. Then, by multiplying this number by adjustment factors based on the quality of the protein, the relevant values for other diets containing lower-quality proteins are generated. For instance, if the optimal amount of protein required by people of a given size, age, and sex is X when the protein in their diet is at its highest quality, then the requirement would be 2X when the protein in their diet is only 50% utilized, 4X when the protein is only 25% utilized, and so on. It is evident from this approach to determining protein requirements that the nutritive quality measure, whatever it may be, must range linearly from a maximum of 100 to a minimum of zero. There are serious questions about the validity of these assumptions raised by recent observations.

### 2.3.5.2. Biological Value (BV)

Protein nutritive value has traditionally been estimated using biological value, as stated by Thomas (1909) and Mitchell (1924, p. 26). It is described as the "percentage of absorbed nitrogen retained in the body," and measuring the biological value and digestibility of dietary protein is a necessary part of a thorough analysis of the protein. After feeding the test protein, the amounts of nitrogen voided when fed a nitrogen-free diet are adjusted for, and the fecal and urine nitrogen levels are measured to determine these results. The percentage of dietary nitrogen absorbed from the intestine is known as true digestibility.

$$Digestability = \frac{I - (F - F_0)}{I} \times 100$$

Therefore, Biological Value is:

$$BV = \frac{I - (F - F_0) - (U - U_0)}{I - (F - F_0)} \times 100$$

where; $I$ is the Nitrogen intake of test protein; $F$ is the Fecal nitrogen; $F_0$ is the Fecal nitrogen on nitrogen-free diet (Metabolic N); $U$ is the Urinary nitrogen; and $U_0$ is the Urinary nitrogen on nitrogen-free diet (Endogenous N).

To avoid extreme weight loss, Mitchell (1926) showed that in practice, the endogenous nitrogen was very similar to that achieved when a little amount of high-quality protein was supplied. Therefore, it was chosen to feed limited amounts of egg protein rather than a nitrogen-free diet. The fundamental premise behind measuring biological value is that endogenous and metabolic nitrogen are constants that may be rightfully deducted from test readings using the formula. There's not much evidence to imply that this isn't always the case. For instance, methionine treatment may significantly reduce the excretion of urine nitrogen in rats and dogs fed a nitrogen-free diet, resulting in a Biological Value of methionine alone that is significantly higher than 100%. Although it hasn't been fully investigated, this might not occur in humans. Furthermore, gelatin's Biological Value was discovered by Mitchell et al. (1927) to be 20%, or 20% as satisfying as the highest-quality proteins. Gelatin by itself cannot sustain animals, hence this must be an overestimation of the true nutritional value. This gap seems to be comparable to that found in NPU values for diets with low intakes of the majority of essential amino acids reported by Bender (1961). The Mitchell technique should be used to determine a protein's total nutritional value (Net Protein Value), which is equal to Biological Value x Digestibility.

### *2.3.5.3. Net Protein Utilization (NPU)*

Similar to biological value, NPU calculates nitrogen retention by comparing the body nitrogen content of animals given a test protein to those fed no protein. "Percentage of the dietary protein retained" is the NPU, which is calculated by dividing this value by the total amount of protein consumed.

NPU = body N content with test protein − body N content with protein-free diet × 100 / N intake

Replica groups of four weanling rats kept in group cages and given either the "protein-free" or the "test" food for 10 days was the process Miller (1963) advocated. The specific benefits of these circumstances need to be proven, as they were selected empirically. Body water measurements have been frequently substituted for body nitrogen measurements because there is a strong association between body nitrogen and body water content in young animals (13–16). In fact, because sample errors are avoided, measuring body water may be more accurate than measuring body nitrogen. It is also far more practical and less expensive. Since "retained nitrogen" is the basis for both NPU and BV calculations, they ought to measure the same thing. However, the denominator in BV is the amount absorbed, while the total protein consumed is the denominator in NPU calculations. Given the quantity of nitrogen lost due to poor digestibility (lack of absorption), BV should be larger than NPU. Although it hasn't been demonstrated, total carcass analysis in weanling rats may provide a more precise estimate of "retained nitrogen" than nitrogen balance tests. It is undoubtedly less tiresome. Measurements of the nitrogen balance must be applied to both large animals and human subjects.

### *2.3.5.4. Amino Acid Score*

Originally, Block and Mitchell (1946) hypothesized that a comparable shortfall of any amino acid would inhibit protein synthesis to the same degree because all amino acids must be present at the location of protein synthesis in enough proportions for protein synthesis to begin. They therefore proposed that it should be possible to determine the nutritive value of a protein by computing the deficit of each essential amino acid in the test protein from the amount in the "ideal protein" if the composition of the "ideal protein" was known, that is, a protein that contained every essential amino acid in sufficient amounts to meet requirements without any excess. The nutritive value would presumably be determined by the "most limiting amino acid," or the one in the largest deficiency. Since the whole egg's protein was known to have a Biological Value that was very near to 100, they actually recommended it as the "ideal" protein. They realized that certain amino acids in egg proteins may exceed necessary levels. If this is the case, the deficits of these in other proteins determined by this process would be artificially elevated. In other words, the estimated nutritional value would be less than the real value. When "amino acid deficits" were computed using egg protein as the standard, Block and Mitchell (1946) compared biological values, which were believed to have been measured appropriately. They discovered a rather high correlation (r =.86), indicating the procedure's general validity. Since then, amino acid scores have been applied extensively. Typically, the "percentage of adequacy" has been computed instead of the deficits that Block and Mitchell (1946) proposed. The 1957 FAO Committee proposed that amino acid scores be determined using an amino acid pattern based on estimations of human amino acid requirements, acknowledging once more that egg proteins may contain different essential amino acids above what is needed. The Food and Nutrition Board's Amino Acid Committee suggested a similar course of action. The second FAO/WHO Expert Group, however, concluded that the previously recommended pattern was inappropriate in certain ways and that there wasn't enough data to say that the nutritional value of the proteins in eggs, cow's milk, or human milk varied. As a result, they proposed that any of these patterns may be regarded as "optimal" to determine amino acid scores. This proposal has caused difficulty in the Amino Acid Score calculation because the amino acid makeup of these three proteins differs significantly. Additionally, they proposed that there may be a relationship between protein quality and the ratio of essential amino acid nitrogen to total nitrogen (E/T). This has caused more confusion because no mechanism for combining this ratio with the Amino Acid Score was offered.

### *2.3.5.5. Protein Efficiency Ratio (PER)*

As previously mentioned, there are numerous ways to show qualitative variations in protein quality. Because of its ease of use, Protein Efficiency Ratio

(PER) has been the most used approach. Young rats fed specific proteins gained little weight and ate little protein, while those fed higher-quality proteins gained greater weight and ingested more protein, according to Osborne, Mendel, and Ferry's (1919) observations. They computed the weight increase per gram of protein consumed, or PER, in an effort to make up for the variation in food intake. It is well known that the amount of protein included in the test diet affects the PER for any given protein. As a result, standardized conditions have been suggested. These consist of using 10 weanling rats per test group, feeding them meals with 9.09% protein (N × 6.25), having a 4-week test period, and having a group in each experiment that receives standardized casein. The average total weight increase divided by the average number of grams of protein consumed yields the PER. It was advised to compute a corrected value using an assumed 2.50 PER of the standardized casein (Corrected PER = 2.50 × PER/PER of reference casein) because PER for the same protein varied throughout laboratories. PER has been harshly criticized as a protein quality metric despite its simplicity. The most frequent objections have been that body composition varies and is not a reliable indicator of nitrogen retention, and that some food protein is needed for the animal's maintenance, but this is not attributed to the protein in the PER measurement.

The main theoretical objections of PER are that it is not directly related to the nutritional value of the protein but rather depends on factors such as weight increase, food intake, protein content, and the quality of the protein in the diet. These have an unclear and complicated relationship. Another drawback of PER is that, even in typical operating environments, it cannot be replicated in other laboratories. The fact that the corrected PER values in the collaborative study (Derse, 1962) displayed greater variations between laboratories than the uncorrected values is noteworthy since it suggests that the correction was inappropriate and unhelpful. It is evident that the PER does not correlate with the nutritional quality of the proteins examined; for instance, a protein with a PER of 1.5 cannot be taken to have 50% of the value of a protein with a PER of 3.0. Therefore, a requirement for Textured Protein Products that states that "the total protein (must have); a Biological Value not less than 70% of casein" is meaningless. Many times, PER can be used to determine whether one protein is superior to another, but since the differences are not proportionate to nutritional content, it is inappropriate to express these differences as percentages.

### *2.3.5.6. Net Protein Ration (NPR)*

The PER has been heavily criticized for not accounting for the amount of protein needed for maintenance because it solely considers weight increase when making calculations. Bender and Doell (1957) proposed that a group of animals on a diet devoid of protein may be included in each test to circumvent this problem. Next, the Net Protein Ratio (NPR) was computed by dividing the total gain (weight gain of the test group plus weight loss of the protein-free group) by the amount of protein consumed. It is clear that this process is the same as NPU if body composition remains constant, with the exception that the arbitrary units used for expression are less helpful than the proportion of protein used. Naturally, the shortcomings are the same as those covered under NPU.

### 2.3.5.7. Relative Nutritive Value (RNV)

The slope of the regression line comparing body protein (or body water) of a standard protein (egg protein or lactalbumin), which is thought to have the highest nutritional value, was compared to that of the test protein in Hegsted et al.'s (1970) slope-ratio assay, which was conducted on rats. This approach thus tests the implicit assumption that the NPU or BV measurements are independent of the amount of protein supplied. The initial assumption was that the regression line should bisect the Y axis at the position indicated by the group given the protein-free diet, just like in the computation of NPU and BV. As was previously mentioned, this does not happen frequently, if at all. In contrast to Miller and Payne's findings (1961), the regression lines above the maintenance level of intake are linear over a wide range of intakes with young, growing rats.

This approach is arguably the most conceptually justifiable of the assays available as an estimate of the protein quality for growth, especially in young growing rats where the requirements for maintenance are low in comparison to those for development. Whether protein quality estimations for growth in young rats are suitable for estimating quality for humans, including young infants, is still a crucial subject. Many proteins will presumably be used by humans more effectively than by developing young rats.

### 2.3.5.8. Nitrogen Balance Index

As was previously mentioned, Allison and Anderson (1945) demonstrated that the slope of the regression line between nitrogen intake and nitrogen balance is the biological value. They also hypothesized that this approach may offer some practical benefits over the standard BV calculation technique. This index and the previously stated Relative Nutritive Value share a similar notion. The validity of this index has to be confirmed because it is becoming more and more evident that nitrogen retention is not linearly related to nitrogen intake in the region of intake below maintenance.

### *2.3.5.9. Tissue Regeneration*

Many methods for restoring weight or particular tissues following protein deficiency have been put forth. It is still to be shown what advantages these particular tests have over, say, young rats' weight gain.

### *2.3.5.10. Microbiological Assays*

The necessary amino acids needed by monogastric mammals are needed by many microorganisms. An easy-to-use and effective method of determining the nutritional value of a protein would be to identify organisms that require the same pattern of amino acids in the same relative amounts. This would allow researchers to observe how these organisms grew in response to varying amounts of proteins or protein hydrolysates. A lot of work has gone into this, and it's evident that some organisms' reactions are similar to those shown in some of the rat tests that have been detailed. The challenges are evident, nevertheless, as the animal tests' shortcomings make them an insufficient foundation for comparison with assays of this type.

### *2.3.5.11. Plasma Amino Acids*

Under certain circumstances, changes in plasma amino acid levels following the consumption of different proteins can provide estimations of the nutritional quality. However, it should be mentioned that each amino acid's range in the plasma of normal animals is somewhat wide. The quantitative interpretation of any changes in the levels measured is severely limited by this variability. Therefore, while this method may be able to determine the limiting amino acid in some proteins, the chances of developing accurate quantitative assays for nutritional quality based on plasma amino acid levels are not encouraging.

A closer look at the many approaches used to assess the quality of proteins indicates that things are not as good as people had thought. It is likely because different essential amino acid shortages develop at different rates that NPU and BV fail to demonstrate the predicted link to the Amino Acid Score. For example, higher than predicted NPUs or BVs are produced by the longer period needed to induce severe lysine insufficiency compared to most other amino acid deficiencies. Thus, the most logically sound way to assess the quality of protein may be to use the Amino Acid Score. This, of course, depends on accurately assessing the requirements for amino acids. But unless more advanced or novel biological techniques for determining protein quality are developed, it will continue to be mostly unjustifiable. Protein quality may be rather accurately tested for growth in young animals, such as rats. The main concern that needs to be addressed is whether or not these metrics are appropriate for a creature that grows relatively slowly, like humans. It seems likely that adult or slowly growing species have comparatively higher-quality proteins than do young growing rats. This suggests that the proportion and quantity of essential amino acids needed for growth and maintenance are different.

## 2.4. VITAMINS

Vitamins are a class of organic compounds with distinct chemical properties that are required in trace amounts for various physiological processes. The word "vitamin" comes from the term "vital amine," which describes a class of substances with particular functions in metabolism. Despite being organic substances, vitamins are not needed for the formation of structural compounds and do not offer energy like other macronutrients do. Nonetheless, they serve as coenzymes or precursors of enzymes in several metabolic processes. While some vitamins can be produced by the rumen and hindgut bacteria or by exposure to sunlight, the majority must be given to the animal by food. Vitamin deficiencies in the diet cause illnesses, lower productivity and animal welfare, and weakened immune systems in animals raised for food. Vitamin needs are quite low in the diet. Megadoses of some vitamins (such as vitamin E) have been added to animal diets recently in an effort to boost immunity and enhance food quality. Based on their solubility, vitamins can be broadly categorized as water- or fat-soluble. Vitamin A, vitamin D, vitamin E, and vitamin K are the fat-soluble vitamins. Vitamin C and other members of the B-complex group are among the water-soluble vitamins.

### 2.4.1. Classification of Vitamins

There are two types of vitamins: fat-soluble and water-soluble. There are 13 vitamins in humans: eight B vitamins and vitamin C are water-soluble, and four fat-soluble (A, D, E, and K). Because water-soluble vitamins are easily dissolved in water and generally eliminated from the body, urine production is a good indicator of vitamin intake. It is crucial to consume them more consistently because they are not as easily stored. Lipids (fats) aid in the absorption of fat-soluble vitamins via the digestive system. The body may get overloaded with vitamins A and D, which could lead to hazardous hypervitaminosis. Malabsorption-related deficiencies in fat-soluble vitamins are especially important in cystic fibrosis patients.

Fat-soluble vitamins:

- Vitamins A, D, E, K;
- Associated with fat during digestion and absorption;

- Storage in the liver, adipose tissue, and excess storage can be toxic for some vitamins (e.g., A and D);
- No daily need;
- Deficiency is very slow.

Water-soluble vitamins:

- Total nine, all B vitamins and vitamin C;
- Soluble in water and excess excreted through urine;
- No storage and less toxic;
- Daily requirement (except vitamin B12);
- Serve as cofactor in biochemical reactions;
- Deficiency is fast.

### *2.4.1.1. Vitamin A*

M. Mori named this vitamin A when he identified it in 1922 as a "fat-soluble factor" found in butter and fish oil. The terms retinol (alcohol), retinal (aldehyde), and retinoic acid (acid form) refer to a group of related chemicals that are collectively referred to as vitamin A. Retinol is the biologically active form of vitamin A among these three compounds. All animals need to consume vitamin A in their diets.

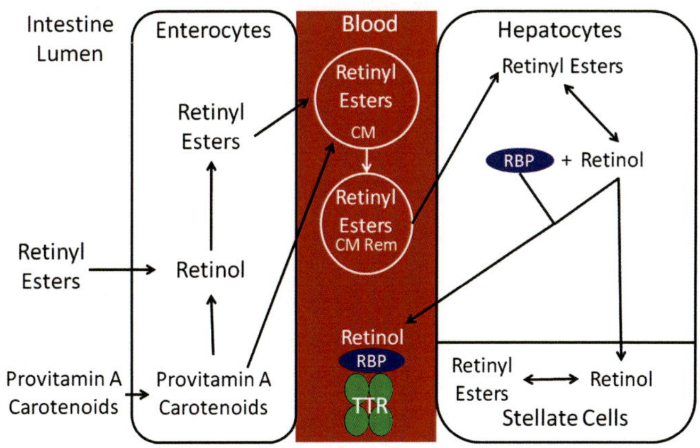


Figure 2.12. *Different forms of vitamin A.*

*Source: https://courses.lumenlearning.com/suny-nutrition/chapter/12-62-vitamin-a-uptake-absorption-transport-storage/.*

The food can supply vitamin A either directly from the vitamin or from its precursor, carotenoids, which are found in plants. The majority of vitamin A in animal feed comes from synthetic sources, which are more affordable to produce. Vitamin A's precursor or plant form is called carotenoids. Carotenoids are pigments (more than 600 varieties) found in plant cells that give foods like carrots, sweet potatoes, and pumpkins their rich orange or yellow color. Carotenoids come in two different forms: xanthophylls and carotenes. Of these, carotenes possess vitamin A activity, particularly β-carotenes. The other types of carotenoids found in plants, called xanthophylls, are responsible for the pigmentation of color but lack vitamin action. Diets for plumage color enrichment (e.g., exotic birds kept in captivity), egg yolk pigmentation, aquaculture feeds, and decorative fish diets are increasingly using these types of carotenoids (Figure 2.12).

### *2.4.1.1.1. Function*

Vitamin A is involved in multiple physiological processes in the body, such as bone formation, vision, reproduction, and epithelial cell maintenance. Epithelial cells cover the skin and mucous membranes found in body cavities, such as the respiratory, urogenital, and digestive systems. It is commonly known that vitamin A has a function in night vision. Rhodopsin, sometimes known as visual purple, is the result of the combination of retinal with opsin, a protein found in the rods of the retina. Because rhodopsin is light-sensitive, the eye can adjust to variations in light levels. Rhodopsin breaks into retinal and opsin when exposed to light. The optic nerve receives the energy that is emitted, which causes vision. Nevertheless, a deficiency in the retina causes ineffective recycling of rhodopsin, which renders rod cells hypersensitive to variations in light and ultimately results in night blindness. Cell differentiation and proliferation require vitamin A. The synthesis of mucoprotein, a barrier that shields cells from bacterial invasions, also requires vitamin A. Vitamin A deficiency can cause epithelial cells to fail to differentiate into mature, mucus-producing cells. Additionally, defective, stratified, keratinized cells can replace normal epithelial cells, making the body more vulnerable to infection. A vitamin A deficiency causes xerophthalmia, a disorder that affects both people and animals. It causes the cornea and conjunctiva of the eye to become dry and irritated, which can lead to infection and cloudiness. Additionally, healthy tooth and bone growth as well as reproductive functions depend on vitamin A. The maintenance of cell membranes and the division of osteoclasts and osteoblasts—bone cells—are related to vitamin A's role in bone formation. Additionally essential for reproductive processes including spermatogenesis and estrus cycles is vitamin A. Carotenoids and vitamin A have the ability to act as antioxidants, shielding cells from oxidative stress. They also play a role in regulating humoral and cell-mediated immune responses in animals.

### *2.4.1.1.2. Metabolism*

Fat in the diet is absorbed and digested along with vitamin A. Vitamin A is found in the food in the form of esters, which are broken down by pancreatic lipase and then combined with lipid micelles. When they get to the microvilli, they are moved to mucosal cells, where they undergo reesterification, become part of the chylomicrons, and are then carried to the lymph for eventual liver storage as retinyl esters. The blood carries the hydrolyzed retinyl esters, which liberate retinol, to the required tissues where it combines with retinol-binding proteins. Within the intestinal mucosal cells, carotenoids divide in two to generate retinal and are reduced to form retinol. However, the bioconversion of carotenoids to retinol varies greatly throughout mammals. About 0.6 μg of β-carotenes equals one IU of vitamin A. Certain animals, including cats, require preformed vitamin A from animal sources because they lack the enzyme needed to convert β-carotene breaking dioxygenase to vitamin A.

### *2.4.1.1.3. Toxicity*

Since vitamin A is a fat-soluble vitamin, prolonged use of it may result in severe side effects. But depending on the species, age, and physiological state, the symptoms will change. Skin thickening and skeletal problems have been linked to hypervitaminosis. Dietary supplementation of vitamin A on production performance of poultry is given in Table 2.4.

Table 2.4. *Vitamin A Function, Deficiency, and Excess Toxicity*

| Parameter | Species | Dietary Supplementation | Effect |
|---|---|---|---|
| | Layers | 9,000 IU/kg | Increased |
| | Layer | 5,000 IU/kg (T5) to 25,000 IU/kg (T25) | No effect |
| | Layer | 16,000 IU/kg | Increased |
| | Broiler | 0 or 3,000 IU/kg | Decreased |
| | Broiler | 5,000 IU/kg | Increased |
| | Layers | 8 mg/kg canthaxanthin | Increased |
| | Broilers chicks | 15,000 IU/kg | Increased |
| | Layer | 16,000 IU/kg | Increased |
| | Growing ducks | Above 2,500 IU/kg | Decreased |
| | Growing ducks | 0 to 2,500 IU/kg | Increased |
| | Broiler | 2,645 to 7,935 IU/kg | Decreased |
| | Broiler | 35,512 and 65,512 IU/kg | Decreased |
| | Broiler | 0 or 3,000 IU/kg | Decreased |

| | | | |
|---|---|---|---|
| Drip loss | Broiler | 10,00,000 IU/kg | Decreased |
| Shear force | Goslings | 21,600 IU/kg | Increased |
| pH | Breeder laying hens | 9,000 IU/kg | Increased |
| pH | Broilers chicks | 40 mg/kg lycopene | Increased chick body weight |
| Redness | Broiler | 15,000 IU/kg | Increased |
| Carcass traits | Broiler | 3,208~6,208 IU/kg | Increased |
| Carcass quality | Broiler | 10,000 IU/kg | Increased |
| Relative liver weight | – | 15,000 IU per kg | Decreased |
| Hatching performance | Broiler | Overdose | Decreased |
| Hatchability | Broiler | 10,000 to 15,000 IU/kg | Decreased |
| Reproductive performance | Layer | 12,000 IU/kg | Increased |
| Hatchability and reproduction | Layers | 16,000 IU/kg | Increased |
| Hatching capacity | Layers | 9,000 IU/kg | Increased |
| – | Layers | 9,000 IU/kg | Increased |
| | Layer | 5,000 to 25,000 IU/kg | No effect |
| | Layer | 2,10,000 and 410,000 IU/kg | Decreased |
| | Layer | 16,000 IU/kg | Increased |
| Egg production | Layer | Overdose | Decreased |
| Egg production | Layers | 24,000 IU/kg | No effect |
| Laying performance | Layers | 9,000 IU/kg | Increased |
| Egg yield | Broiler | 10,000 to 15,000 IU/kg | Decreased |
| Egg yield | Layers | 24,000 IU/kg | No effect |
| Efficiency of vitamin A transfer from diet to egg | Layer | 8,000 IU/kg | No effect |
| Egg production | Layer | 210,000 and 410,000 IU/kg | Decreased |
| Number of eggs produced | Layer | 16,000 IU/kg | Increased |
| Egg size | Broiler | 10,000 to 15,000 IU/kg | Decreased |
| Egg production | Layer | 5,000 to 25,000 IU/kg | No effect |
| Egg yolk mass and egg weight | Layers | 8 mg/kg canthaxanthin | Increased |

| | | | |
|---|---|---|---|
| Egg yolk CX concentration | Broiler breeders | 6 mg/kg canthaxanthin | Increased |

*Source: Khan, R. U., Khan, A., Naz, S., Ullah, Q., Puvača, N., Laudadio, V., Mazzei, D., Seidavi, A., Ayasan, T., & Tufarelli, V. (2023). Pros and cons of dietary vitamin A and its precursors in poultry health and production: A comprehensive review. Antioxidants, 12(5), 1131. https://doi.org/10.3390/antiox12051131.*

## *2.4.1.2. Vitamin D*

One of the group sterol compounds found in vitamin D controls the body's metabolism of calcium and phosphorus. Vitamin D, sometimes known as the "sunshine" vitamin, is created when sterols in plants and animal skin are exposed to radiation. Ergocalciferol (vitamin D2, activated plant form) and cholecalciferol (vitamin D3, activated animal form) are the two main forms of vitamin D (Figure 2.13).

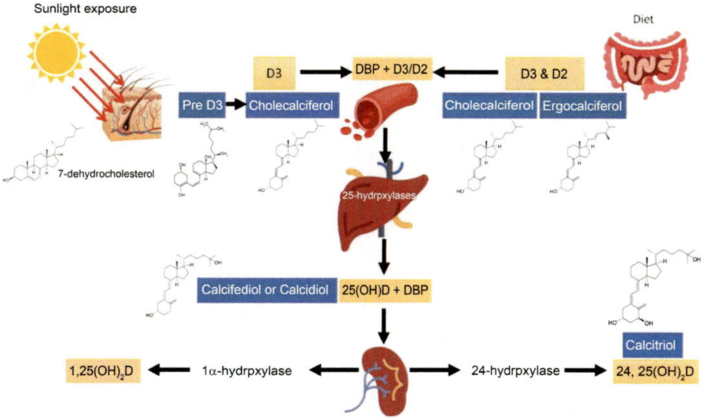


Figure 2.13. *Plant and animal forms of vitamin D.*

*Source: Dominguez, L. J., Farruggia, M., Veronese, N., & Barbagallo, M. (2021). Vitamin D sources, metabolism, and deficiency: Available compounds and guidelines for its treatment. Metabolites, 11(4), 255. https://doi.org/10.3390/metabo11040255.*

In plants, ergocalciferol (vitamin D2) is not generated in live plant cells but rather when the plant is exposed to sunlight following harvest or injury. Forage treated with sunlight and hay provides grazing ruminant animals with a strong supply of vitamin D. Animals confined to enclosed spaces, such as those found in contemporary commercial pig and poultry farms, will need vitamin D. Other omnivores and carnivores depend on the activated animal form of vitamin D3, or cholecalciferol. Vitamin D2 can be changed into vitamin D3 in most mammals. In poultry, conversion efficiency is extremely low. The type of vitamin D that most animals can benefit nutritionally is cholecalciferol, or vitamin D3. When cholesterol is exposed to UV radiation and changes into 7-dehydrocholesterol in the skin, the body produces vitamin $D_3$. It is transferred from the epidermis to the liver, where it is hydroxylated to become 25-hydroxycholecalciferol, for it to become active. The blood carries this substance to the kidneys, where it undergoes further hydroxylation to generate 1,25-hydroxycholecalciferol, or calcitriol, the most metabolically active form of vitamin D.

### *2.4.1.2.1. Functions*

Vitamin D is regarded as a hormone since it is produced by the body and because it regulates calcium and phosphorus balance. Furthermore, parathyroid hormones and the gastrointestinal system, kidneys, and bones collaborate with vitamin D to maintain blood calcium balance and promote bone calcification. By modifying the food absorption of calcium from the gastrointestinal system and the release of calcium from bone, normal blood calcium levels can be attained. For calcium and phosphorus to be properly absorbed from the intestine, proteins that bind to calcium are required. Vitamin D promotes the production of calcium-binding proteins in the gastrointestinal system, which makes it possible for the body to absorb calcium and phosphorus from food. In cases of hypocalcemia, or low blood calcium levels, parathyroid hormones (PTH) indirectly promote calcium absorption from the gut and resorption from the kidney tubules and bone by promoting the synthesis of vitamin D. Through its interaction with PTH to mobilize calcium from bone and its induction of increased P resorption in the kidneys, vitamin D also influences proper bone development and calcification. Together, the kidneys, intestines, and bones cooperate with vitamin D to keep blood calcium and vitamin P levels at an ideal level, which is necessary for healthy bone mineralization. Vitamin D deficiency causes aberrant skeletal growth and poor bone mineralization, which in turn causes rickets in young animals and osteomalacia in growing animals. Insufficient bone calcification causes lameness, twisted legs, and spontaneous long bone fractures in each of these cases. A few minutes in the sun can avoid a vitamin D deficit, albeit the amount of sunlight needed varies depending on skin pigmentation. Animals with lighter skin tones need less sunlight than those with darker skin tones. Comparably, because of the characteristics of their high-elevation habitat and exposure to sun radiation, animals like llamas have higher requirements. Moreover, vitamin D is utilized to treat dairy cows with milk fever. Egg yolks and organ meats are two nutritious food items that are suitable for feeding pets. Cod liver oil is another dense source of information. Vitamin

D (cholecalciferol) is added as a vitamin D supplement to the diet of animals (Table 2.5).

**Table 2.5.** Effects of Vitamin $D_3$ on the Female Reproductive System of Farm Animals and Birds

| Tissue | Species | Cell Type | Treatment | Effect | References |
|---|---|---|---|---|---|
| Ovary | Goat | Gc | 1, 10, 100 nM | ↑ proliferation | Yao, et al., (2017)., Yao et al., (2020). |
| | Goat | Gc | 10 nM | ↓ reactive oxygen species | Yao et al., (2020). |
| | Goat | Gc | VDR silencing | ↑ apoptosis | Yao et al., (2020). |
| | Goat | Gc | 10 nM | ↑ P4 and E2 | Yao, et al., (2017). |
| | Pig | SF, MF | 10, 50 ng/mL | ↑ P4 | Grzesiaket al., (2021). |
| | Pig | Gc | 100 nM | ↓ P4 | Honget al., (2016). |
| | Pig | Gc | 100 ng/mL | → P4 | Smolikova et al., (2013). |
| | Pig | Gc | 100 ng/mL | ↑ E2 | Smolikova et al., (2013). |
| | Pig | Gc | 100 ng/mL | ↑ E2 | Honget al., (2016). |
| | Pig | SF | 100 ng/mL | ↑ E2 | Grzesiaket al., (2021). |
| | Pig | MF | 1–100 ng/mL | ↑ E2 | Grzesiaket al., (2021). |
| | Pig | SF, MF, LF | 1–100 ng/mL | → T | Grzesiaket al., (2021). |
| | Chicken | Gc | 10, 100 nM | ↑ proliferation | Wojtusik & Johnson, (2012). |
| | Chicken | Gc | 10, 100 nM | ↓ *AMH* mRNA | Wojtusik & Johnson, (2012). |
| | Chicken | Gc | 100 nM | ↑ *FSHR* mRNA | Wojtusik & Johnson, (2012). |

| | | | | | |
|---|---|---|---|---|---|
| | Chicken | Gc | 100 nM | ↑ *Kit ligand* mRNA | Kundu et al., (2012) |
| | Chicken | – | Vitamin $D_3$-deficient diet | ↓ E2 and P4 | Ruschkowski & Hart, (1992). |
| Uterus | Pig | M | 10, 50 ng/mL | ↑ E2 | Grzesiak et al., (2019). |
| | Pig | E | 2–200 µM | ↑ implantation-related genes | Jang, et al., (2017). |

*Abbreviations: AMH: Anti-Müellerian hormone; FSHR: follicle-stimulating hormone receptor; Gc: granulosa cells; E: endometrium; E2: estradiol; LF: large follicles; M: myometrium; MF: medium follicles; P4: progesterone; SF: small follicles; and T: testosterone.*

*Source: Grzesiak, M., Tchurzyk, M., Socha, M., Sechman, A., & Hrabia, A. (2022). An overview of the current known and unknown roles of vitamin $D_3$ in the female reproductive system: Lessons from farm animals, birds, and fish. International Journal of Molecular Sciences, 23(22), 14137. https://doi.org/10.3390/ijms232214137.*

### 2.4.1.3. Vitamin E

Tocopherols and tocotrienols are a class of chemically related chemicals that are collectively referred to as vitamin E.

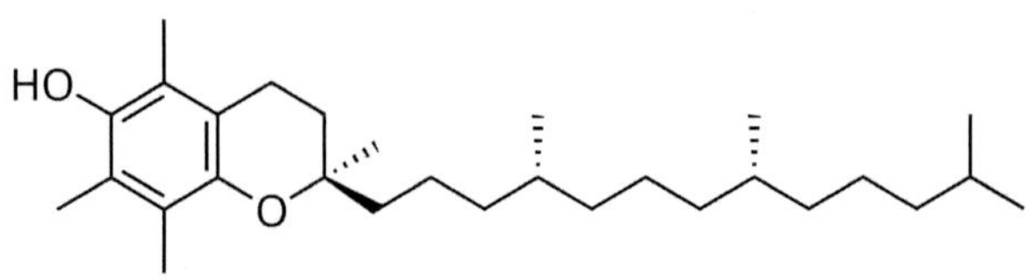


Figure 2.14. *Vitamin E structure.*

*Source: By Calvero – Own work (made with ChemDraw), Public Domain, https://commons.wikimedia.org/w/index.php?curid=1556458.*

The most active biological form of vitamin E among its many isomers, α-tocopherol, is the one that is added to animal diets. α-, β-, γ-, and δ tocotrienols and β-, β-, γ-, and δ tocopherol are other isomers that have fewer biological effects. DL-α-tocopheryl acetate is the most common form of vitamin E found in stores. For vitamin E, one milligram of all-rac-α-tocopherol acetate equals one International Unit (IU). The most biologically active form of vitamin E in animals is RRR-α-tocopherol, the only stereoisomer of α-tocopherol found in nature (Figure 2.14).

### *2.4.1.3.1. Functions*

The body uses vitamin E as a biological chain-breaking antioxidant to prevent oxidative damage from free radicals and other products of lipid oxidation, as well as to shield cells and tissues from these agents. By acting as a free radical scavenger and donating electrons from the molecule's hydroxyl group, vitamin E stops fats from oxidizing (antioxidant effect). Lipid peroxidation damages unsaturated lipids in cell membranes, which compromises the integrity of the cell and the structural membrane. The development of such peroxidized chemicals in prepared feeds can lower the food's organoleptic and sensory quality while also diminishing palatability, rancidity, and nutritional value. It can also have an impact on animal health. Vitamin E can shield other nutrients like proteins and vitamin A from oxidative stress as well as lipids. Because of these functions, the amount of polyunsaturated fatty acids, the degree of peroxidative damage, and other external stresses all affect the amount of vitamin E in a diet. Additionally, vitamin E has a sparing effect on selenium, a mineral needed as a cofactor for the enzyme glutathione peroxidase, which lowers lipid peroxides. Lipid peroxide inactivation shields the cell membrane from more harm. Selenium is spared by vitamin E because it stops the polyunsaturated fatty acids in cell membranes from oxidizing. Plant oils and egg yolks are two food sources that are high in vitamin E, depending on the diet of the hens. Because vitamin E is easily destroyed, it is important to store prepared feed properly—away from heat and light—to preserve vitamin E levels and stop oxidative alterations to fat.

### *2.4.1.3.2. Deficiency*

White muscle disease, exudative diathesis, and encephalomalacia can result from a vitamin E deficit. Degeneration of the heart and skeletal muscle fibers is the cause of white muscle disease, which results in heart failure and an early death. Leaky capillaries in the breast muscle produce exudative diathesis in hens. In both situations, vitamin E or selenium treatment will be effective. But the sole cure for encephalomalacia, or crazy chick illness, is vitamin E.

### ***2.4.1.3.3. Toxicity***

Since vitamin E is the least harmful of the fat-soluble vitamins, animals (beef, cattle, and chicken) receive large amounts of it in their meals to improve the nutritional content, visual appeal, and lipid stability of their food (Table 2.6).

**Table 2.6.** *Diseases Associated with Vitamin E Deficiency in Different Animal Species*

| Disorder | Animal Model | Compromised Organ/Tissue |
|---|---|---|
| Immune deficiency | Chick, pig | Mononuclear phagocyte system |
| Myopathic disorders | Rabbit, duck, lamb, calf, turkey, chicken | Heart, skeletal muscles, gizzard |
| Reproductive dysfunction | – | – |
| Embryonic apoptosis | Hen, turkey, cow | Embryonic circulatory system |
| Infertility (male) | Rooster, rabbit | Testes |
| Kidney, pancreas, liver, brain, blood | – | – |
| Necrobiosis | Pig | Liver |
| Erythrocyte hemolysis | Chick, calf | Red blood cells |
| Hypoproteinemia | Chick, turkey | Ricin |
| Cerebral softening | Chick, duckling | Encephalon |
| Hemorrhagic diathesis | Chick, turkey | Vascular system |
| Nephrosis | Mink, rat | Renal tubular |
| Yellow fat disease | Pig | Adipose tissue |

*Source: Shastak, Y., Obermueller-Jevic, U., & Pelletier, W. (2023). A century of vitamin E: Early milestones and future directions in animal nutrition. Agriculture, 13(8), 1526. https://doi.org/10.3390/agriculture13081526.*

### *2.4.1.4. Vitamin K*

Quinones are a class of chemicals that are included in vitamin K. Phylloquinones, or vitamin K1, are present in green plants, while menaquinones, or vitamin K2, are produced by bacteria in the hindgut. The gastrointestinal (GI) tract can easily absorb vitamin K in conjunction with fat. Before being used, vitamins K1 and K3 are changed by the liver to K2. Menaquinones are the type of vitamin K that is biologically active. The most prevalent type of vitamin K found in animal diets is menadione (vitamin K3, synthetic form).

## 2.4.1.4.1. Functions

The blood-clotting protein prothrombin requires vitamin K to be synthesized. Several proteins, including thromboplastin, prothrombin, fibrinogen, and fibrin, are required for the blood-clotting process. Because the enzymes required for these activities depend on vitamin K, deficiencies in this vitamin can cause hemorrhages, extended bleeding times, and/or failure in the creation of fibrin clots. Subcutaneous hemorrhages frequently occur over the surface of the body, giving the skin a blotchy, bluish look.

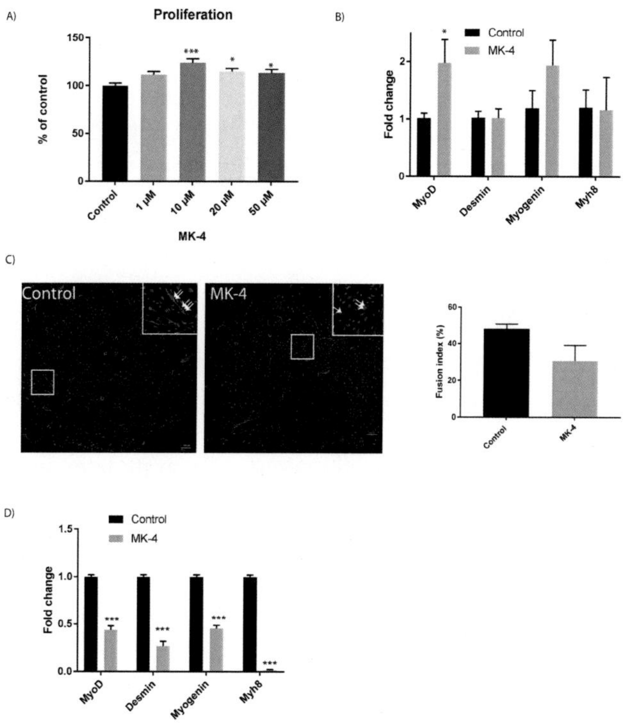


Figure 2.15. *Vitamin K function: Its effect on bovine cells.*

*Source: Rønning, S. B., Pedersen, M. E., Berg, R. S., Kirkhus, B., & Rødbotten, R. (2018). Vitamin K2 improves proliferation and migration of bovine skeletal muscle cells in vitro. PLoS ONE, 13(4), e0195432. https://doi.org/10.1371/journal.pone.0195432.*

This may also have financial ramifications for animals raised for food, which could lower carcass quality or result in condemnation. Most animals can obtain the necessary vitamin K from intestinal bacteria through coprophagy, or eating excrement, or absorption from the hindgut. However, the amount of vitamin K that is available in an animal's food might be restricted by antibiotic therapy or animal husbandry methods such as confinement housing or keeping animals in cages or pens with wire floors. Since sulfa medications are antagonists of vitamin K, certain coccidiostats that contain them may result in vitamin K insufficiency. The mold that grows on weather-damaged sweet clover hay or silage contains dicoumarol, a substance that shares many structural similarities with vitamin K. Vitamin K is competitively inhibited by dicoumarol. When calves eat moldy sweet clover hay or silage, they become deficient in vitamin K, which causes internal bleeding and ultimately results in death. Warfarin is a rat poison that also inhibits vitamin K and causes anticoagulation. It also inhibits vitamin K competitively. Since the primary ingredient in these rodenticides, warfarin is an anticoagulant and can induce bleeding and hemorrhaging, vitamin K is frequently given to pets that have had rodenticide poisoning. Even at higher concentrations, derivatives of menaquinone and phylloquinone are safe. However, animals who get sustained high dosages of menadione develop anemia and other problems (Figure 2.15).

### 2.4.1.5. B Vitamins

The original classification of the B vitamins (also known as B complex vitamins) was based on their comparable metabolic roles. The nine chemically unrelated organic compounds serve as coenzymes, or metabolic catalysts, for the production of blood cells in animal bodies, cellular maintenance, and energy metabolism pathways. Thiamine, riboflavin, niacin, pyridoxine, pantothenic acid, biotin, folic acid, cobalamin, and choline are the B vitamins covered in this section. Microbial synthesis satisfies the needs of ruminant animals and herbivores, while daily supplementation is necessary in monogastric animals like pigs and poultry. Additionally vulnerable to loss during feed processing are B complex vitamins.

#### 2.4.1.5.1. Thiamine

One pyrimidine molecule linked to one thiazole molecule makes up thiamine. Since thiamin was the first vitamin to be discovered, it is also known as vitamin B1. Thiamine is a part of the thiamin pyrophosphate (TPP) enzyme, which is engaged in multiple important events in pathways that produce energy. Within cells, dietary thiamine is transformed into TPP to take part in the mechanism that produces energy. TPP is necessary for oxidation decarboxylation reactions like pyruvate dehydrogenase and α-ketoglutarate dehydrogenase. In the pentose cycle, it is also employed in transketolase processes to produce nicotinamide adenine dinucleotide phosphate (NADPH). The amount of carbohydrates in an animal's diet affects how much thiamine it needs because of its function in the metabolism of carbohydrates. Thiamine can be found in cereal grains. However, feed processing can destroy thiamine content since thiamine is heat labile. The energy amount of the diet is related to the thiamin requirement (0.5 mg/1,000 kcal diet).

- **Deficiency:** Because beriberi, a disorder of the nervous system, is a common thiamin deficit, thiamin also has a special significance in neurophysiology. Another common sign of thiamin insufficiency in chicks is polyneuritis. A thiamine deficiency can result from several substances that have chemical structures similar to thiamine and that can act as antagonists. The enzyme thiaminase, which is found in raw fish and perennial bracken ferns, breaks down thiamin, resulting in a deficit that leads to Chastek paralysis, a neurological illness named for a farmer who noticed a similar ailment in silver foxes. Thiaminase is denatured by heat treatment, which stops the issue. Amprolium, also known as coccidiostat, inhibits TPP activation and may result in thiamine shortage.

#### *2.4.1.5.2. Riboflavin*

Flavin, or yellow color, and ribose, or sugar, are the sources of the name riboflavin. Vitamin B2, riboflavin, is rapidly damaged by light, yet it is rather heat-stable. In the body, riboflavin is a part of two distinct coenzymes: flavin adenine dinucleotide (FAD) and flavin mononucleotide (FMN). The tricarboxylic acid (TCA) cycle, oxidation, and electron transport chain are three processes in which both of these enzymes are engaged and which release energy from proteins, lipids, and carbohydrates.

- **Deficiency:** Similar to the majority of B vitamins, insufficient amounts cause juvenile animals to grow less. In young animals, low riboflavin diets can result in diarrhea, hair loss, and sores in the corners of the mouth in addition to anorexia.

#### *2.4.1.5.3. Niacin*

Pyridine 3-carboxylic acid and its derivatives demonstrating the nutritional action of nicotinic acid are generally referred to as niacin. Niacin is a component of two crucial coenzymes: NADPH and nicotinamide adenine dinucleotide (NAD). These coenzymes function as hydrogen carriers in a number of critical metabolic pathways, including the TCA cycle and oxidative phosphorylation, which derive energy from proteins, lipids, and carbohydrates. One such system is carbohydrate metabolism or glycolysis. The niacin included in cereal grains is bound and not physiologically accessible to animals when fed in their diets.

For instance, niacytin, one of the binding forms of niacin found in wheat, is not physiologically accessible. Niacin is firmly bound by the niacinogen found in maize, preventing absorption. Animal-based sources of niacin are widely available. Apart from cats, the majority of mammals can produce niacin through synthesis from the necessary amino acid tryptophan, in addition to food sources. Thus, tryptophan levels can influence the amount of niacin needed. Nonetheless, diets deficient in tryptophan are typically also low in niacin. Niacin is something that cats should get entirely from their food.

- **Deficiency:** In canines, this results in a disorder known as black tongue disease; in hens, it causes spectacled eyes or poor eye feathering. A diet of 10–90 mg/kg is necessary. A common sign of niacin deficiency in humans, pellagra (thick skin, dermatitis) is linked to low income and a bad diet high in grains and meatless foods.

### *2.4.1.5.4. Pyridoxine*

There are three different types of pyridoxines: plant-based pyridoxine, animal-based pyridoxal, and animal-based pyridoxamine. The physiologically active form of the coenzyme pyridoxal 5-phosphate is pyridoxal. Pyridoxal 5-phosphate is involved in a broad range of metabolic events, including transamination, deamination, and decarboxylation, the majority of which are related to the metabolism of amino acids. Additionally, hemoglobin production and the conversion of tryptophan to niacin depend on pyridoxal phosphate. Niacin deficiency can be brought on by a vitamin B6 deficiency. As a result, the symptoms of these two vitamin deficiencies are comparable.

- **Deficiency:** Convulsions and a weakened immunological response are among the symptoms. Dietary requirements range from 1–3 mg/kg and are related to the amount of protein consumed.

### *2.4.1.5.5. Pantothenic Acid*

There contains pantothenic acid in every bodily tissue. The name of the vitamin comes from the Greek word pan, which means "all" or "everywhere." It was discovered that pantothenic acid is a component of coenzyme A, which is necessary for the acetylation of many different molecules in the metabolism of energy. To enter the citric acid cycle and synthesize steroids, two-C fragments from lipids, amino acids, and carbohydrates must be formed. This process requires CoA. In rare instances, a deficiency in this vitamin causes pigs to grow at a slower rate and develop an irregular gait known as "goose-stepping," which is caused by degeneration of the nerves. Rough coats, anorexia, and decreased output are other indicators of insufficiency. The most popular way to include the vitamin in diets is as Ca salt.

### *2.4.1.5.6. Biotin*

Rats fed raw egg whites develop skin lesions and hair loss, which are treated by a liver-derived protective factor. This chemical was originally known as vitamin H because it preserved haut, which is German for "skin." Year 1936 saw the isolation of biotin, a yeast growth factor, from egg yolk. As a prosthetic group, biotin forms a peptide bond with the enzyme's lysine to generate biocytin. Biocytin is then used as a cofactor in carboxylase reactions, including pyruvate carboxylase (the first step in gluconeogenesis) and acetyl CoA carboxylase (the first step in lipogenesis). Biotin plays a critical role in the metabolism of lipids and carbohydrates through these vital metabolic pathways. In reactions where carbon chains are extended, biotin serves as a transporter of carbon dioxide (carbon fixation).

- **Deficiency:** Deficit in biotin is uncommon. It results in hair loss and dermatitis. The antivitamin avidin binds to biotin, preventing it from being absorbed and digested, which is typically the cause of this rather than a shortage of biotin in the diet. A great source of biotin is eggs. But avidin is found in egg whites. Cooking, on the other hand, denatures avidin and releases the biotin for absorption. Since hindgut bacteria create biotin, deficiency symptoms may be observed in pigs housed in pens with slotted floors and little to no access to fecal matter. It has been demonstrated that in hens fed wheat-based diets, a deficiency in biotin results in a disease known as footpad dermatitis. A diet containing 0.1–0.3 mg of biotin per kg is required (dry basis). Animals receiving antibiotic medication that results in a reduction of the bacterial population can require an additional biotin supply.

### *2.4.1.5.7. Folic Acid*

The general word for folic acid and its related chemicals is foliacin. Tetrahydrofolic acid is the term for the active form of folacin in the body. Tetrahydrofolic acid is mostly produced in the liver from dietary sources of folacin. The conversion of folacin to tetrahydrofolic acid is facilitated by vitamin B12. Tetrahydrofolic acid serves as a means of transportation for monocarbons. The synthesis of purines, pyrimidines, glycine, serine, and creatine requires tetrahydrofolic acid. DNA synthesis and thus cell replication depend on the production of both purines and pyrimidines.

- **Deficiency:** All mitotically active cells are impacted by folic acid deficiency, which reduces DNA and cell replication. Hematopoietic cells and all epithelial cells are among them. Megaloblastic anemia is the result, as rapidly dividing cells are the most impacted. The most common vitamin deficit in humans is folic acid insufficiency. A folic acid shortage during pregnancy may affect up to one-third of pregnant women globally. There is a strong link between folic acid and vitamin B12; folic acid shortage will arise from vitamin B12 deficiency. Including antimicrobials raises the risk of folate insufficiency. A diet containing 0.25 mg of folic acid per kilogram is required.

#### *2.4.1.5.8. Cobalamin*

The final B vitamin, cobalamin (Vitamin B12), was identified in 1948. The vitamin is called cyanocobalamin, and the coenzyme form is called deoxyadenosyl cobalamin. The active site of vitamin B12 is a trace element mineral called cobalt, which makes it special. Furthermore, it is the only vitamin that is exclusively produced by microbes. Cobalamin participates in the transfer of single carbon units throughout a variety of metabolic events, just like folic acid does. For the methyl transfer pathway in the metabolism of fats and carbohydrates as well as the formation of myelin, folic acid functions as a coenzyme in many enzyme systems. In ruminant animals, cobalamin is necessary for the oxidation of propionic acid. When it comes to absorbing vitamin B12, the stomach is crucial. The acidity and pepsin produced by the stomach enable the dietary source of tightly bound vitamin B12 to be released. Additionally, the stomach secretes a particular binding glycoprotein known as intrinsic factor. The compound of intrinsic factor and vitamin B12 makes its way to the ileum, where it is absorbed into a portal vein. The ileum needs calcium in order to absorb vitamin B12. A lack of glycoprotein may result in a vitamin B12 shortage. The signs of insufficiency and folic acid deficiency are very similar. For nonruminant diets, the need is incredibly low: 5–50 μg/kg. Only ruminants need cobalt; cobalamin is made by the rumen microorganisms.

- **Deficiency:** This is comparable to folic acid deficiency, which can lead to neurological problems and anemia. Nutrient absorption is hampered by deficiency in folacin or vitamin B12. Shorter villi and alterations to the intestinal epithelial cells are seen. Reduced growth and loss of appetite are found in livestock species. There is no need because all B vitamins are synthesized by rumen microorganisms in ruminants. But in order for them to synthesize vitamin B12, they need cobalt. Cobalamin requirements range from 5 to 50 μg/kg feed (nonruminants). All ruminants require is cobalt.

### *2.4.1.6. Vitamin C*

The discovery that consuming lemon juice helps prevent scurvy dates back to 1747. In 1933, ascorbic acid, or vitamin C, was acknowledged as a vitamin. The structure of ascorbic acid is similar to that of monosaccharide carbohydrates. The majority of animal species and plants make it from glucose. No specific coenzyme form is found. The hydroxylation processes of proline and lysine in the creation of collagen, elastin, and neurotransmitters (epinephrine and norepinephrine) require ascorbic acid. Normal bone development requires collagen. It lessens oxidative stress by acting as an antioxidant as well. With the exception of primates and guinea pigs, all animals can synthesize ascorbic acid from glucose. Therefore, cattle species are not necessary. Collagen production requires proline and OH-proline.

#### *2.4.1.6.1. Deficiency*

This leads to scurvy, a disease that initially appeared in sailors at sea and affects people with poor bone development, anemia, and reduced wound healing. All mammals, with the exception of primates and guinea pigs, often do not exhibit any signs of insufficiency. There is no set daily requirement for livestock animals (Table 2.7).

Table 2.7. Water-Soluble Vitamins, Coenzymes/Cofactors, and Functions

| **Water-Soluble Essential Vitamins** | | | |
|---|---|---|---|
| **Vitamin** | **Function** | **Deficiencies Can Lead to** | **Sources** |
| Vitamin $B_1$ (Thiamine) | Needed by the body to process lipids, proteins, and carbohydrates Coenzyme removes $CO_2$ from organic compounds. | Muscle weakness, Beriberi: reduced heart function, CNS problems. | Milk, meat, dried beans, whole grains. |
| Vitamin $B_2$ (Riboflavin) | Takes an active role in metabolism, aiding in the conversion of food to energy (FAD and FMN). | Cracks or sores on the outer surface of the lips (cheliosis); inflammation and redness of the tongue; moist, scaly skin inflammation (seborrheic dermatitis). | Meat, eggs, enriched grains, vegetables. |

| | | | |
|---|---|---|---|
| Vitamin $B_3$ (Niacin) | Used by the body to release energy from carbohydrates and to process alcohol; required for the synthesis of sex hormones; component of coenzyme $NAD^+$ and $NADP^+$. | Pellagra, which can result in dermatitis, diarrhea, dementia, and death. | Meat, eggs, grains, nuts, potatoes. |
| Vitamin $B_5$ (Pantothenic acid) | Assists in producing energy from foods (lipids, in particular); component of coenzyme A. | Fatigue, poor coordination, retarded growth, numbness, tingling of hands and feet. | Meat, whole grains, milk, fruits, vegetables. |
| Vitamin $B_6$ (Pyridoxine) | The principal vitamin for processing amino acids and lipids; also helps convert nutrients into energy. | Irritability, depression, confusion, mouth sores or ulcers, anemia, muscular twitching. | Meat, dairy products, whole grains, orange juice. |
| Vitamin $B_7$ (Biotin) | Used in energy and amino acid metabolism, fat synthesis, and fat breakdown; helps the body use blood sugar. | Hair loss, dermatitis, depression, numbness, and tingling in the extremities; neuromuscular disorders. | Meat, eggs, legumes, and other vegetables. |
| Vitamin $B_9$ (Folic acid) | Assists the normal development of cells, especially during fetal development; helps metabolize nucleic and amino acids. | Deficiency during pregnancy is associated with birth defects, such as neural tube defects and anemia. | Leafy green vegetables, whole wheat, fruits, nuts, legumes. |
| Vitamin $B_{12}$ (Cobalamin) | Maintains healthy nervous system and assists with blood cell formation; coenzyme in nucleic acid metabolism. | Anemia, neurological disorders, numbness, loss of balance. | Meat, eggs, animal products. |
| Vitamin C (Ascorbic acid) | Helps maintain connective tissue: bone, cartilage, and dentin; boosts the immune system. | Scurvy, which results in bleeding, hair and tooth loss; joint pain and swelling; delayed wound healing. | Citrus fruits, broccoli, tomatoes, red sweet bell peppers. |

*Source: https://courses.lumenlearning.com/wm-biology2/chapter/essential-nutrients/.*

## 2.5. MINERALS

The physiological and metabolic activities of an animal's body depend on minerals, which are inorganic substances. Mineral matter constitutes around 4% of an animal's body weight and is essential for maintaining life and health. More than any other single class of nutrients, minerals are critical to every biological process in the body. These processes include gene expression and regulation, which control enzyme systems and cell activity, vitamin function, osmotic balance, detoxification, immunity, cell membrane function, acid-base balance and regulation, as well as growth and structural support (such as bones). Scientific literature lists around 21 important minerals. However, this chapter will focus only on those minerals that are unequivocally linked to various animal nutrition issues.

Minerals are divided into two categories—macro- and micro-(trace) minerals—based on dietary requirements, not their importance to physiological processes. Macrominerals are required in significant amounts (>0.01%) in the diet and are found in noticeable concentrations in the animal's body. Examples include calcium, phosphorus, magnesium, sulfur, and electrolytes (sodium, potassium, and chloride). Microminerals, on the other hand, are required in parts per million, milligrams, micrograms, or trace amounts (<0.01%). Microminerals covered include manganese, zinc, iron, copper, selenium, iodine, cobalt, molybdenum, and chromium. These elements must be supplied in a diet as salts, combined with other elements, and cannot be added in their elemental forms (e.g., NaCl, $CaCO_3$, $MnSO_4$).

### 2.5.1. Calcium and Phosphorus

#### *2.5.1.1. Functions*

In the mammalian body, phosphorus (P) and calcium (Ca) both serve as structural elements. Hydroxyapatite, a substance found in bones and teeth, contains about 99% of the calcium and 80% of the phosphorus found in mammalian bodies. The remaining 1% of calcium is found in cellular fluids, where it is used for a variety of metabolic and physiological processes, including blood coagulation, the maintenance of nerve impulse and cell permeability, the activation of specific enzymes, muscular contraction, and the function of activators of ion channels. The body's soft tissues contain phosphorus, which plays a crucial role in phosphorylation events as well as cellular oxidative pathways for energy consumption. For instance, phosphorus is a part of adenosine triphosphate (ATP), a phosphorylated molecule that is essential to energy metabolism. Phosphorylated pentose sugars are also present in deoxyribonucleic acid (DNA) and ribonucleic acid (RNA). In addition, phosphorus is a component of the phospholipids found in cell membranes, which are necessary for cellular fluidity and nutrient transfer. Consequently, phosphorus and calcium are essential for several metabolic processes that support animal existence. Grain cereals contain a lot of phosphorus. Phytate or phytic acid, on the other hand, are bonded forms of P found in cereal grains. P's availability from these constrained sources ranges from 20% to 60%. The phrase "available P" is frequently used to refer to unbound forms of P in the diet of monogastric animals because these animals do not have the enzyme phytase to release them from the bound form. Microbial phytase, an enzyme that may break and release P, is produced by ruminant animals.

### *2.5.1.2. Regulation of Blood Calcium Levels*

The maintenance of a constant level of circulating blood plasma calcium is known as homeostasis, and it is a tightly regulated physiological regulation in the body. As indicated below, parathyroid hormone (PTH), calcitonin, and active forms of vitamin D are responsible for this. The parathyroid gland releases parathyroid hormone (PTH) in response to low blood calcium (hypocalcemia), which increases the absorption of calcium from food through the gastrointestinal (GI) tract, increases P excretion into urine, and increases the synthesis of active forms of vitamin D in the kidneys (Table 2.8).

**Table 2.8.** More Daily Milk and Better Milk Composition in Cows Fed Dicalcium Phosphate

| **Parameters** | **Supplementation of DCP at different level in % DM/d** | | | | | |
|---|---|---|---|---|---|---|
| | **0** | **0.3** | **0.6** | **0.9** | **SE** | **p-value** |
| Initial milk yield* | 9.95$^{c}$ | 10.75$^{b}$ | 10.77$^{ab}$ | 11.75$^{a}$ | 0.35 | 0.0001 |
| Av. Daily milk yield | 10.04$^{c}$ | 12.74$^{b}$ | 12.87$^{ab}$ | 14.08$^{a}$ | 0.47 | 0.0161 |
| Difference | 0.09$^{c}$ | 1.99$^{b}$ | 2.10$^{ab}$ | 2.33$^{a}$ | 0.01 | 0.0182 |
| Milk composition | | | | | | |
| Fat | 3.62$^{d}$ | 4.15$^{c}$ | 4.36$^{b}$ | 4.73$^{a}$ | 0.15 | 0.0096 |
| Protein | 2.85$^{c}$ | 3.48$^{b}$ | 3.72$^{ab}$ | 4.15$^{a}$ | 0.18 | 0.0127 |
| Lactose | 3.60$^{b}$ | 4.23$^{ab}$ | 4.30$^{a}$ | 4.30$^{a}$ | 0.18 | 0.0211 |
| Total solids | 12.10 | 11.50 | 11.60 | 12.57 | 0.71 | 0.7078 |

*Source: Wondater, W. B., & Ayanie, T. D. (2023). Supplementing dairy feed by di-calcium phosphate and effect on dry matter intake, digestibility, milk composition, and blood mineral balances in crossbred dairy cows. PLoS ONE, 18(11), e0282879. https://doi.org/10.1371/journal.pone.0282879.*

By boosting osteoclast activity, vitamin D, in conjunction with parathyroid hormone (PTH), enhances the mobilization of calcium from bone, resulting in an elevation of blood calcium levels to normal ranges. In response to elevated blood calcium levels (hypercalcemia), the thyroid gland's parafollicular cells release calcitonin, a hormone that counteracts PTH by reducing calcium mobilization from bones and restoring normal blood calcium levels.

In the diet, calcium and phosphorus have a significant reciprocal relationship, with the calcium-to-phosphorus ratio being a crucial factor. Excess dietary calcium can form insoluble compounds with phosphorus, leading to reduced phosphorus absorption. Similarly, a diet high in phosphorus or phytate phosphorus (P) can hinder calcium absorption. While cereal grains contain large amounts of phosphorus, much of it is bound as phytate phosphorus (30–60%), which reduces the absorption of other nutrients by forming complexes resistant to digestive enzymes.

For small animals, the optimum Ca:P ratio is 1:1; for large animals, it is 2:1. Calcium phosphorus imbalance can be caused by feeding meals with an incorrect Ca:P ratio or by supplementing feeds with high concentrations of one of these minerals. These issues have an impact on animal welfare, productivity, and skeletal health and may result in financial loss.

### *2.5.1.2. Calcium Phosphorus Deficiency or Imbalance Disorder in Animals*

Minerals, especially calcium and phosphorus, are stored in bone. Unbalanced levels of calcium and phosphorus therefore cause anatomical abnormalities in animals and poor eggshell quality in laying hens. In animals raised for food, an excess or imbalance of calcium and phosphorus is linked to a number of bone growth issues. A disease known as rickets affects young, growing animals and is brought on by normal growth in the organic matrix but inadequate mineralization. Animals who are adults and eat a diet low in calcium develop osteomalacia. Brittle, demineralized bones result from an excessive loss of calcium in the bone. Loss of bone's organic matrix and mineralization combine to cause osteoporosis. Due to incorrect calcification, bones become soft and frequently distorted in both osteomalacia and rickets. Ca shortage can cause significant alterations in animals that grow quickly, such as pigs and chickens, because of the fast turnover of skeletal minerals in these animals. Large animals, like sheep and cows, take longer to exhibit the symptoms of deficiencies. Osteomalacia may be accompanied by lameness, limb weakness, irregular gait, and spontaneous fractures. In every instance of a Ca deficit or Ca-P imbalance, there is a decrease in the amount of bone ash. Hypocalcemia from a severe calcium deficiency might result in tetany and seizures. A well-known case of hypocalcemia and calcium tetany in dairy cows is milk fever, also known as parturient paresis. The animal experiences a reduction in body temperature, tetany, and finally falls, bending its head over its flank. The decreased blood calcium levels are the cause of this. The goal of treatment is to raise the calcium level in the blood by injecting calcium salts such as calcium chloride, calcium lactate, or calcium gluconate intravenously. To improve the absorption of calcium, a high-dose vitamin D injection should be administered five days before calving. Because of the high calcium need of breastfeeding combined with hormonal insufficiency, milk fever in high-producing dairy cows always occurs within the first 24 hours postpartum. Normally, the high requirement for milk calcium is satisfied by using the calcium minerals found in bone.

On the other hand, PTH in particular controls the hormonal regulation of bone mineral mobilization. A low-calcium diet should be given at least 14 days before calving to "prime" or boost endocrine activity. This will enhance PTH secretion, which will increase calcium mobilization from bones when lactation commences. The topic of electrolyte balance is covered under electrolytes and is crucial in preventing milk fever. The majority of the time, cows with milk fever recover quickly after receiving intravenous calcium. Cage layer weariness is a condition that typically affects high-producing young chickens at the peak egg production phase (>35 weeks of age), much like milk fever. A lot of calcium is needed for egg laying in order for the eggshell to develop. Leg weakness results from increased bone mobilization caused by insufficient calcium. Affected birds may withdraw to a corner of the cage, exhibit reluctance to move, or lay eggs with soft shells or deformities. For them to produce eggs, laying hens need a minimum of 3.3 g of calcium each day, which is far more than other animals require. It is generally advised that hen diets have sources of calcium and bigger particle sizes, which improve retention in the intestine. One of the biggest welfare problems older chickens have after laying is osteoporosis, which results in fractured bones and weak legs. The development and growth of bones are influenced by the Ca:P ratio. Pets fed homemade meat-based meals or animals fed grain-based diets with subpar hay frequently have excess P and low Ca. Large breeds of dogs fed extra Ca-supplemented diets and young horses fed high-energy diets are at risk for developmental bone-related problems. It is advised to follow a diet that has a 1:1 to 2:1 Ca:P ratio along with sluggish bone formation. When fed meat-only diets instead of the meat-and-bone diets they take in the wild, huge cats (tigers, cheetahs) maintained in zoos have been found to have similar symptoms associated with Ca-P imbalance. Low Ca and high P levels in the blood are caused by a low Ca-P ratio. PTH secretion increases in response to such circumstances, promoting the excretion of P in the urine and the mobilization of calcium from bone. Long-term dietary imbalances in chronic situations cause nutritional secondary hyperparathyroidism. The affected animals exhibit lameness, unwillingness to move, joint pain, edema, demineralized bones, and loss of bone mass. P-deficient soil is found in many tropical regions of the world, and animals that graze in these locations frequently acquire a perverted appetite and strange chewing and feeding habits known as pica. Elevated fluoride disrupts the digestion and absorption of P. Animals suffering from P shortage will eventually become polluted by fluorine gas from industrial sources in pastures (Table 2.9).

**Table 2.9.** *Phosphorus: Blood Metabolites of Young Growing Male Goats Fed an N- and/or P-Reduced Diet*

| Item[1] | N+/P+ | N-/P+ | N+/P- | N-/P- | SEM | *p*-value | | |
|---|---|---|---|---|---|---|---|---|
| | | | | | | N-reduction | P-reduction | Interaction |
| *n* | 7 | 7 | 7 | 7 | | | | |
| Calcium (mmol/L)* | 3.16[a] | 3.11[a] | 4.34[b] | 3.55[a] | 0.24 | 0.002 | <0.001 | 0.005 |
| Calcidiol (nmol/L) | 47.71 | 53.57 | 56.86 | 50.71 | 6.38 | 0.965 | 0.335 | 0.072 |
| Calcitriol (pg/mL) | 18.46 | 14.57 | 13.39 | 18.76 | 3.48 | 0.673 | 0.801 | 0.014 |
| cAMP (pmol/mL) | 21.51 | 20.59 | 36.63 | 17.63‖ | 11.24 | 0.090 | 0.291 | 0.121 |
| IGF1 (ng/mL) | 243.8[a] | 193.8[a] | 123.8[b] | 116.3[b] | 32.38 | 0.090 | <0.001 | 0.202 |
| GH (ng/mL) | 5.84 | 6.22 | 4.90 | 5.31 | 1.43 | 0.580 | 0.209 | 0.987 |
| Glucose (mg/dL)* | 68.00 | 67.86 | 67.43 | 66.57 | 4.19 | 0.814 | 0.662 | 0.866 |
| Magnesium (mmol/L) | 0.95[a] | 0.72[b] | 1.09[a,c] | 1.01[a] | 0.09 | 0.003 | <0.001 | 0.120 |
| Phosphate (mmol/L)* | 2.09[a] | 2.00[a] | 0.70[b] | 1.06[b,c] | 0.26 | 0.301 | <0.001 | 0.099 |
| T3 (nmol/L) | 1.83 | 2.04 | 1.94 | 1.94 | 0.28 | 0.464 | 0.943 | 0.446 |
| Urea (mmol/L)* | 6.76[a] | 1.16[b] | 6.85[a] | 2.22[b,c] | 0.83 | <0.001 | 0.180 | 0.250 |
| 24,25$(OH)_2D_3$ (ng/mL) | 8.58 | 8.68 | 4.88 | 7.29 | 2.47 | 0.318 | 0.050 | 0.361 |

(Mean values with their pooled standard errors; $n = 6–7$ animals).
[1]IGF1, Insulin-like growth factor 1; T3, Triiodthyronine 3.
[a,b,c]Mean values within a row with unlike superscript letters were significantly different; Tukey's multiple comparisons test ($p < 0.05$).
*Data already published in Weber et al. (23).
‖ $n = 6$; due to the outlier test performed.

*Source: Zillinger, L. S., Hustedt, K., Schnepel, N., Hirche, F., Schmicke, M., Stangl, G. I., & Muscher-Banse, A. S. (2024). Effects of dietary nitrogen and/or phosphorus reduction on mineral homeostasis and regulatory mechanisms in young goats. Frontiers in Veterinary Science, 11, 1375329. https://doi.org/10.3389/fvets.2024.1375329.*

## 2.5.2. Magnesium

The body contains magnesium as phosphates, which is the third most abundant element, and as carbonates in bone, liver, and skeletal muscle cells. Magnesium has structural roles in the skeletal system and is necessary for the activation of many enzymes that split and transport phosphatases within cells. Magnesium is a cation found in intracellular fluid that is involved in the metabolism of proteins and carbohydrates. Magnesium is essential for both nerve impulse transmission and muscular contraction, together with calcium, sodium, and potassium. The ileum is where most dietary magnesium is absorbed. For Mg to be absorbed, no carrier is required. Magnesium absorption is unaffected by vitamin D. Uncertainty surrounds the homeostatic regulation of magnesium in blood and tissue, and PTH promotes the release of magnesium from bone. Serum magnesium levels are lowered, and urine excretion is increased in nutritional secondary hyperparathyroidism.

#### *2.5.2.1. Deficiency*

Food sources contain large amounts of magnesium. Grass tetany is a typical issue in livestock that grazes. Another name for it is "wheat grass poisoning." It is especially common in animals that graze in springtime on native pastures or verdant meadows rich in grain forages. It is mainly common on meadows with wheat grass. It has been connected to an increase in springtime trans aconitate in the green pasture. Mg is bound by trans aconitate, which causes Mg insufficiency. Tetany of the muscles, head retraction, stumbling, convulsion, and acute sensitivity to touch or sound are among the symptoms. Magnesium absorption is inhibited by potassium and nitrogen. Richly fertilized pastures typically have high concentrations of N and K; animals that graze on these pastures are more prone to grass tetany. A magnesium solution is injected intravenously, magnesium is fed from various sources, pastures are rotated, and dry forages are provided as treatments.

### 2.5.3. Sulfur

The structural element of sulfur (S) is found in skin, hair, wool, feathers, cartilage, and connective tissue. The body needs sulfur mostly as a part of organic substances that include sulfur. These include the hormone insulin, the anticoagulant heparin, the chondroitin sulfate, and mucopolysaccharide, which is present in cartilage. Methionine, cysteine, and cystine are the three amino acids that also include sulfur. Amino acids containing S make up the majority of S in the body. It is commonly advised that sheep's diets for wool growth and birds' diets during periods of rapid feather growth contain an amino acid with a high S content. Enzymes that serve as antioxidants, such as glutathione peroxidase, also contain sulfur. S is also a part of two B vitamins that are important in the metabolism of fats and carbohydrates: biotin and thiamin. S is crucial for energy metabolism as well because it is a part of coenzyme A.

#### *2.5.3.1. Deficiency*

Sulfur insufficiency can result in weight gain and decreased growth of feathers and wool. Dietary inorganic S absorption is incredibly weak. Amino acids that contain S can provide organic S, which satisfies the S requirement. When nonprotein nitrogen is added to the diet, S supplementation may aid in microbial protein production and weight gain in sheep.

#### *2.5.3.2. Toxicity*

S poisoning is not a realistic issue due to the extremely low intestinal absorption.

### 2.5.4. Electrolytes

#### *2.5.4.1. Sodium, Potassium, and Chlorine*

The animal body is kept electrically neutral by electrolytes, which are dissolved compounds that carry an electrical charge. The difference between total cation and anion intake and excretion determines the acid-base balance. Since these three minerals are electrolytes and aid in maintaining ionic equilibrium as well as the viability of cells, they are covered together in this section together with sodium (Na), potassium (K), and chlorine (Cl). The electrolytes are essential for preserving the osmotic pressure in intra- and extracellular fluids, cell membrane signal transductions, and the acid-base balance (pH maintenance in blood and tissue). Electrolyte normal ratios are extremely stable between species. The body of an animal contains regulatory mechanisms to manage the levels of these minerals. They must, however, be consumed every day because they cannot be preserved. All animals receive common salt (NaCl) in their diets, and grazing animals are allowed to choose from it at will. Other trace elements, such as iodized salt or trace mineralized salt, are also delivered by salt. It is a routine procedure to add 0.3% to 0.5% of salt to the diets of pigs and poultry. The primary extracellular cation that is present in blood and extracellular environments is sodium ($Na^+$). Together with other ions, sodium keeps cells permeable so that nutrients can actively pass through membranes. A significant portion of the body's basal metabolic rate is regulated by the sodium pump, or Na-pump, which also manages electrolyte homeostasis. Additionally, neuronal impulse transmission and muscle contraction depend on sodium. Animal diets contain sodium in the form of sodium chloride (NaCl). The main cation that is present in higher concentrations inside the cells (intracellular fluid) is potassium (K). Osmotic force, which keeps fluid volume constant, is produced by ionized K inside the cells. Additionally, cellular potassium participates in a number of enzymatic processes. It is necessary to maintain potassium balance for the heart muscle to contract normally. The negatively charged anion that balances the effect of the positively charged cations (K and Na) is called chloride. Two-thirds of the anion in extracellular fluid that is involved in controlling osmotic pressure is chlorine (Cl). Hydrochloric acid, which is required for the activation of gastric enzymes and the start of protein digestion in the stomach, is formed when chlorine is also needed. Animal diets that contain NaCl provide them with chloride.

### 2.5.4.2. Deficiency

These three nutrients are often found in reasonable amounts in typical diets, and deficiencies are uncommon.

### 2.5.4.3. Toxicity

Normal electrolyte excretion is regulated by the kidneys. Toxicology is therefore quite uncommon unless it results from renal illnesses, hydration restriction, or excessive salinity in the water. Excessive sodium consumption, while linked to hypertension in humans, has not been documented in animals.

### 2.5.4.4. Dietary Electrolyte Balance

The maintenance of animal health, welfare, and productivity in food animal production has sparked a growing interest in the electrolyte balance. Animal health and productivity can be negatively impacted by acidosis or alkalosis caused by changes in the acid-base balance. Dietary electrolyte balance is often reported as Na + K – Cl (meq/kg). The ideal equilibrium for chickens is 250 meq/kg, while the range for pigs is 100–200 meq/kg dry matter (DM). Dietary electrolyte imbalance has been linked to poor growth, which affects appetite, anomalies of the legs, such as tibial dyschondroplasia (slipped tendon), and poor growth, production, and welfare of chickens. Maintaining electrolyte balance is crucial for ruminant animals to avoid acidosis and alkalosis. In dairy cattle feeding, dietary cation-anion difference is typically used to lower the incidence of milk fever. In dairy cattle, prepartum alkalosis may make milk fever more common, whereas acidosis may shield against it. Forage-rich prepartum diets are likewise high in K, which may impair the cow's capacity to maintain calcium homeostasis and result in milk fever. Diets that lower blood pH can raise blood calcium levels, which will lessen milk fever (Table 2.10).

**Table 2.10.** *Minerals and Their Function in the Human Body*

| Vitamin | Function | Deficiencies can Lead to | Sources |
|---|---|---|---|
| Calcium* | Needed for muscle and neuron function; heart health; builds bone and supports synthesis and function of blood cells; nerve function. | Osteoporosis, rickets, muscle spasms, impaired growth. | Milk, yogurt, fish, green leafy vegetables, legumes. |
| Chlorine* | Needed for production of hydrochloric acid (HCl) in the stomach and nerve function; osmotic balance. | Muscle cramps, mood disturbances, reduced appetite. | Table salt. |
| Copper (trace amounts) | Required component of many redox enzymes, including cytochrome c oxidase; cofactor for hemoglobin synthesis. | Copper deficiency is rare. | Liver, oysters, cocoa, chocolate, sesame, nuts. |
| Iodine | Required for the synthesis of thyroid hormones. | Goiter | Seafood, iodized salt, dairy products. |
| Iron | Required for many proteins and enzymes, notably hemoglobin, to prevent anemia. | Anemia, which causes poor concentration, fatigue, and poor immune function. | Red meat, leafy green vegetables, fish (tuna, salmon), eggs, dried fruits, beans, whole grains. |
| Magnesium* | Required co-factor for ATP formation; bone formation; normal membrane functions; muscle function. | Mood disturbances, muscle spasms. | Whole grains, leafy green vegetables. |
| Manganese (trace amounts) | A cofactor in enzyme functions; trace amounts are required. | Manganese deficiency is rare. | Common in most foods. |

| | | | |
|---|---|---|---|
| Molybdenum (trace amounts) | Acts as a cofactor for three essential enzymes in humans: sulfite oxidase, xanthine oxidase, and aldehyde oxidase. | Molybdenum deficiency is rare. | – |
| Phosphorus* | A component of bones and teeth; helps regulate acid-base balance; nucleotide synthesis. | Weakness, bone abnormalities, calcium loss. | Milk, hard cheese, whole grains, meats. |
| Potassium* | Vital for muscles, heart, and nerve function. | Cardiac rhythm disturbance, muscle weakness. | Legumes, potato skin, tomatoes, bananas. |
| Selenium (trace amounts) | A cofactor essential to activity of antioxidant enzymes like glutathione peroxidase; trace amounts are required. | Selenium deficiency is rare. | Common in most foods. |
| Sodium* | Systemic electrolyte required for many functions; acid-base balance; water balance; nerve function. | Muscle cramps, fatigue, reduced appetite. | Table salt |
| Zinc (trace amounts) | Required for several enzymes such as carboxypeptidase, liver alcohol dehydrogenase, and carbonic anhydrase. | Anemia, poor wound healing, can lead to short stature. | Common in most foods. |

**Greater than 200 mg/day required.*

*Source: https://courses.lumenlearning.com/wm-biology2/chapter/essential-nutrients/.*

## 2.5.5. Microminerals

Based on their dietary requirements, macro- and microminerals differ from one another. Microminerals serve as organic chemical constituents or enzyme activators and are needed in trace amounts (<0.01%, milligrams or microgrammes). This chapter will cover the following microminerals: iodine, cobalt, manganese, zinc, iron, copper, selenium, and molybdenum.

### *2.5.5.1. Manganese (Mn)*

#### *2.5.5.1.1. Functions*

Animals require the trace mineral manganese (Mn) in their diets. Mn is found throughout the animal body, although it is primarily in the liver and bones. The synthesis and upkeep of the mucopolysaccharide in the organic matrix of the bone depend on manganese. As a result, Mn is necessary for healthy bone development. Animals lacking in Mn therefore grow their tendons normally, but their bones grow slowly or abnormally. This causes symptoms including bent calf in young ruminants and perosis (slipped tendon) in chicks. Additionally, manganese is a crucial cofactor for a large number of enzymes that catalyze the metabolism of proteins, fats, and carbohydrates. Mn is mostly found in the mitochondria, where it activates several metal-enzyme complexes that control the metabolism of carbohydrates, including pyruvate carboxylase. Manganese's involvement in the production of fatty acids and cholesterol makes it a cofactor in lipid metabolism as well. Manganese is absorbed from the diet very slowly—less than 10% of total consumption. High dietary Ca or P levels prevent the absorption of Mn. Using transferrin as a carrier, manganese is taken up from the digestive system as Mn 2+, oxidized to become Mn 3+, and then delivered to tissues. A diet high in Mn can cause iron insufficiency.

#### *2.5.5.1.2. Deficiency*

Manganese shortage and default in mucopolysaccharide production are linked to numerous skeletal deformities. Pigs, sheep, goats, and cattle have been documented to exhibit lameness, shortened, and bent legs, and swollen joints. Mn deficiency causes perosis, or slipping tendon, in poultry. The gastrocnemius tendon will slip from its condyle, the tibiotarsal joint will be malformed, and long bones will bend in affected birds. Disorders connected to the bones have been found in cattle fed range lupine. This is due to ingredients in lupine that prevent Mn from being absorbed properly, leading to a deficit. Mn deficiency has been linked to reproductive issues including delayed estrus, poor conception, smaller and less habitable litters in large mammals, and decreased hatchability in birds. Manganese poisoning is quite uncommon. Absorption of Fe is reduced by excess Mn.

### *2.5.5.2. Zinc (Zn)*

The animal body has a large distribution of zinc (Zn). The liver, bones, and animal hides like fur, feathers, skin, and wool all contain high levels of zinc. In the animal body, zinc functions as a cofactor or component (metalloenzyme) in over 100 different enzyme systems. These include enzymes involved in the production of nucleic acids, proteins, and carbohydrates (e.g., DNA and RNA polymerases). The distribution of related enzymes in tissue is closely correlated with the concentration of zinc in that tissue. Because zinc is an ingredient in insulin, it aids in the metabolism of carbohydrates. Zinc is also necessary for the creation of retinol binding proteins and plays a key role in T-cell activity in immunity and reproduction. Zinc absorption varies from 5% to 40% of consumption and is influenced by multiple factors. Low molecular weight binding protein metallothionein has a high affinity for binding zinc and is involved in the metabolism of zinc as well as the transport of zinc from intestinal mucosa cells to plasma. Zinc is bound and trapped inside mucosal cells by metallothionein, which is synthesized in response to high zinc levels. High dietary Ca has a negative impact on zinc absorption, which is exacerbated by the presence of phytate. Dietary phytate chelates with zinc, reducing the animals' access to zinc as well as other minerals like P. Zinc must be absorbed using a common carrier that iron, copper, and zinc share. Consequently, zinc absorption is hampered by high iron. Zinc is lost in stools as mucosal cells shed. A zinc deficit affects cell division and replication. As a result, tissues that grow quickly, like the skin, gastrointestinal system, and reproductive system, are most impacted. Zinc insufficiency is linked to disorders affecting the skin or feathers since zinc is mostly found in bodily coverings like skin, hair, wool, and feathers. Parakeratosis, often known as severe dermatitis, is a zinc deficiency syndrome that results in dry, scaly, cracked skin and poor feathering in chickens. A zinc shortage can cause defective or delayed wound healing because of its involvement in T-cell and immune processes. Elevated calcium and phytate both hinder zinc absorption, which in turn causes zinc insufficiency. Because cereal grains and soybean meal have a high phytic acid content, feeding them to animals increases their zinc requirements.

### *2.5.5.3. Iron (Fe)*

All of the animal's cells contain iron, however, the majority of the iron in the body is found in the protein molecules myoglobin (>4%) and hemoglobin (>65%). Red blood cells include a complex protein called hemoglobin, which is made up of a protein called hemoglobin and a haem group called porphyrin that contains ferrous ($Fe^{2+}$) iron. The production of respiratory pigments, or hemoglobin, which are necessary for transferring oxygen from the lungs to tissues, is the metabolic need for iron. Additionally, iron is a cofactor for a number of metalloenzymes, including catalases, peroxidases, respiratory pigments (myoglobin and hemoglobin), and cytochromes. Dietary iron is provided as the iron bonded to organic molecules, such as hemoglobin, or as inorganic ions, such as ferric or ferrous iron. Ferrous ($Fe^{2+}$) iron is the state in which nonhaem iron is absorbed most often. In the intestine, ferric iron is converted to ferrous iron. The form of iron that is being absorbed is Fe++ or ferrous iron. Due to the body's need for the mineral, the kind of food ingested, and the digestive environment, the duodenum absorbs iron poorly. Inorganic iron in the ferrous form is more easily absorbed than iron in the ferric state, so acidic conditions in the intestine enhance iron absorption. Compared to non-haem iron from plant sources, organic haem iron from animal tissues, such as meat, which contains hemoglobin and myoglobin, is more readily absorbed. Increased absorption occurs during phases of active growth and gestation due to low body reserves and rising metabolic demands. Dietary components such as phytates, tannins, and other divalent metals like zinc, magnesium, and copper can inhibit iron absorption, as they compete for the same binding proteins. Before they can be transferred, the ferrous iron needs to change into ferric iron (Fe+++). Ceruloplasmin, an enzyme containing copper, is needed for this. The main iron-transporting protein in blood is called transferrin, and it contains ferric iron. Iron can be stored as the iron-containing protein ferritin after it enters the enterocyte or it can be moved into the plasma where it attaches to the transport protein transferrin, which is carried through the plasma. Hemosiderin, an insoluble form, and ferritin, a soluble form, are the two other proteins that bind iron in tissues. The body's primary storage locations are the spleen, liver, and bone marrow. Iron is efficiently conserved by most animals; thus, loss is limited unless it is from blood loss from an injury, surgery, parturition, parasite infections, or other causes. Hypochromic (less hemoglobin) and microcytic (smaller cell) anemia, as well as stunted growth, are the results of iron deficiency. These can be linked to either induced iron insufficiency, like cotton pelt in mink, or plain iron deficiency, which is frequent in young pigs (piglet anemia). Iron deficiency in sow milk is exacerbated by competition amongst littermates, fast piglet growth, and inadequate placental maternal transfer, which results in anemia in young pigs. The formaldehyde in the Pacific Hake binds to iron and prevents it from being absorbed, which is what causes cotton pelt in mink. Animals suffering from these deficiencies can be given injections of organic iron, or iron dextran.

### *2.5.5.4. Copper (Cu)*

Hematopoiesis, the process of making red blood cells, requires copper (Cu). Cu and iron metabolism are hence closely associated. Copper is an element that is utilized by many enzyme systems. This includes the lysyl oxidase required for the crosslinking of collagen and elastin. Insufficient crosslinking may result in significant vascular rupture and compromised bone matrices. A component of cytochrome C oxidase, which is involved in the production of ATP and the transport of electrons, is copper. Ceruloplasmin, a plasma protein, is attached to the majority of the copper contained in blood. This Cu-dependent protein serves as a Cu carrier and is required for the binding of plasma iron to transferrin. Additionally, copper is a part of the antioxidant enzyme superoxide dismutase, which neutralizes free radicals and guards against cell death and membrane damage. Tyrosinase, an enzyme, converts the amino acid tyrosine to the pigment melanin; this process requires copper. Insufficient copper can result in ineffective melanin production and decreased pigmentation, which can alter coat color and cause wool (steely wool) to lose its crimp. Ruminant animals' immunity has been demonstrated to be improved by Cu supplementation. Similar to iron, copper is absorbed by an animal according to its needs. The absorption is facilitated by the cysteine-rich protein metallothionein. Cu is utilized by the liver to make ceruloplasmin and other essential proteins after being absorbed, primarily in the duodenum. Cu is also complexed with plasma protein albumin after absorption. By binding to zinc, phytate promotes copper absorption whereas zinc inhibits it. There is a relationship between copper and molybdenum in ruminants. Because excess molybdenum binds to copper and forms an insoluble compound in blood, it can lead to copper shortage. Cu absorption is inhibited by ascorbic acid. Sheep's limited capacity to eliminate copper in bile makes them susceptible to copper poisoning. Hemolysis of red blood cells is caused by copper poisoning. When the liver cells become saturated with copper, oxidative damage results. RBC damage results from the massive amounts of Cu released into the blood during the destruction of liver cells. Hemolysis results in chocolate-colored blood, reddish urine, and kidneys that are metallic green in color. Some dog breeds have an inherited impairment of copper metabolism that results in copper toxicosis.

### *2.5.5.5. Selenium (Se)*

Glutathione peroxidase, an enzyme that breaks down lipid peroxides produced during lipid oxidation, contains selenium. Se and vitamin E have the same ability to stop polyunsaturated fatty acids in cell membranes from peroxiding, preserving the integrity of the cell. As a result, selenium and vitamin E have a mitigating influence on each other's micronutrient requirements. In muscle and blood, Se is also a part of other seleno-proteins. Sperm midpieces need selenium protein. As deiodinase, which changes the thyroid hormone thyroxine into its metabolically active form, triiodothyronine, Se is also involved in the functioning of the thyroid gland. Amino acids that contain sulfur play a significant role in Se metabolism. In the creation of S-containing amino acids, microbes in the rumen substitute selenium (Se) with sulfur (S). These amino acids are then absorbed in the duodenum. Se is kept in storage as selenium and cysteocystine. Different regions of the US and the world have poisonous or Se-deficient soils, which affects the amount of Se in the grains and forages grown there. In all animals, nutritional muscular dystrophy (also known as exudative diathesis or white muscle disease) is caused by Se deficiency. Affected animals exhibit stiffness, trouble moving, and white streaks on their skeletal and cardiac muscles. The calcium salts that are deposited in the deteriorating muscle tissue are what give the skeletal and cardiac muscles their white hue. Chickens with damaged cell membranes experience exudative diathesis, in which cellular fluid seeps into bodily cavities beneath the skin and produces circumstances akin to ascitic. Vitamin E and selenium (Se) can be used to treat symptoms of selenium insufficiency. Retained placenta in dairy cattle may be a result of subclinical Se insufficiency. It has been demonstrated that supplementing animal diets with selenium (Se) improves antioxidant status and lipid stability by lowering lipid peroxidation. Se is needed in extremely small amounts. There is a very small margin between deficiency and toxicity. Five parts per million is harmful, and 0.02 parts per million is needed. Alkali illness is brought on by Se toxicity. Alkali disease-affected animals exhibit aberrant hoof and hair growth, hair loss, and cracked and broken hooves. Se poisoning also results in acute blind staggers, which are brought on by injury to the central nervous system. At this point, the mechanism of action of Se toxicity is unknown. Giving animals a diet high in protein or including inorganic sulfate in their food can help prevent toxicity. Pellets can be given to the rumen for feeding. In an effort to improve the amount of organic Se in the diets of animals foraging in soils poor in Se, forages have recently been fertilized with Se.

### *2.5.5.6. Cobalt (Co)*

One of the components of vitamin B12 is cobalt (Co). There is a large distribution of cobalt throughout tissues, including the kidneys, bones, and liver. It's unclear what forms it takes in tissues other than being a component of vitamin B12. The symptoms of cobalt deficiency are similar to those of vitamin B12 deficiency because cobalt and B12 are closely associated as chelated minerals. Vitamin B12 production in the rumen is decreased when cobalt levels in the feed are low. Animals that are ruminant have a high cobalt need. This is because of their poor absorption capacity and ineffective production of vitamin B12. Emaciated ruminant animals with low levels of cobalt and vitamin B12 are unable to metabolize volatile fatty acids, or propionic acid, for energy production. As a result, the animals have elevated blood levels of propionate and decreased appetites. As propionate is the precursor of blood glucose, hypoglycemia will occur in the impacted animals. Different regions of the world have soils deficit in cobalt, which results in low cobalt levels in the forages that grazing ruminants eat. Ruminants lacking in cobalt are fed dense pellets of the metal orally. These pellets settle in the rumen and give rumen microorganisms the cobalt they need to synthesize vitamin B12. Because inorganic copper is absorbed so slowly from the gastrointestinal (GI) tract, poisoning is rare.

### *2.5.5.7. Iodine (I)*

As a component of the thyroid gland hormones triiodothyronine and thyroxin (tetra iodothyronine), iodine has only one recognized use. The thyroid gland produces tetra iodothyronine, which is then released into the tissues where it transforms into triiodothyronine, the active form. Thyroglobulin, a protein that contains iodine, is the precursor to thyroxine. Thyroxine controls the basal metabolic rate and promotes oxidative cellular activities. The organs with the next largest concentration of I are the thyroid gland, the stomach, the intestine, the mammary glands, and the skin. The thyroid gland is the main organ involved in I metabolism. The thyroid gland contains more than 80% of the iodine in the body. Thyroid-stimulating hormone (TSH), which is released by the anterior pituitary gland, increases the thyroid's absorption of iron. I is mostly kept in the thyroid gland as thyroglobulin, a glycoprotein. Reduced control of the basal metabolic rate (BMR) results from an I deficiency. Animals lacking in vitamin I have tissues that use less oxygen, and growth rates and gonadal activity are correlated with a lower basal metabolic rate. The fur and skin of these animals become brittle and dry. Abortion, stillbirths, or irregular estrus in females and declining quality of semen in males are linked to reproductive issues. Creatinine deficiency manifests as various skeletal abnormalities, skin lesions, and growth failure in young animals. The disease known as goiter is caused by enlarged thyroid glands. The thyroid gland's attempt to produce more thyroxin in response to TSH stimulation is what caused the enlargement. In reaction to decreased thyroxine synthesis, TSH is released. When there is not enough thyroxine to prevent the secretion of TSH, the thyroid gland becomes hyperactive and grows larger (hypertrophy). Animals that consume feeds containing goitrogens, which are chemicals that obstruct the iodination process in the manufacture of thyroxin, may develop goiter. Animals fed such foods may develop an induced I deficit. Goitrogens, or thyroid inhibitors, are found in many plants. The goitrogenic action of plants in the cabbage family, including rapeseed, kale, turnip, and Brassica forages, is noteworthy. It requires between 0.2 and 0.3 parts per million. Chronic long-term high I intake decreases thyroid I uptake and produces toxic symptoms known as hyperthyroidism. All thyroid functions are disrupted by excess I, which raises BMR, pulse rate, and levels of anxiety and excitability.

### *2.5.5.8. Molybdenum (Mo)*

Apart from the microminerals that have been examined, there exist multiple additional elements that exhibit favorable impacts on the growth, immunity, and well-being of animals. Among them are chromium and molybdenum. Molybdenum (Mo) is a cofactor for nitrogenase and xanthine oxidase. Mo is applied to pastures as fertilizer. Mo poisoning is frequently observed, whereas Mo deficiency is uncommon. A surplus of Mo causes a shortage in Cu because it binds Cu in the blood to create an insoluble complex and prevents Cu absorption. It has been determined that chromium (Cr) is a necessary food for animals. It is known that Cr plays a part in both the metabolism of glucose and cells' capacity to absorb it. Supplementing with Cr has been found to boost immunity and lower respiratory illness in cattle, and it is utilized as a feed ingredient in swine nutrition to reduce carcass fat.

## 2.6. WATER

Water is ranked second in importance to oxygen in terms of necessity for life. Water is a macronutrient that is required in higher concentrations than any other item that is consumed orally. Water comes from drinking it, from metabolic water (generated when proteins, lipids, and carbs are broken down into carbon dioxide and water), and from the water that appears as moisture in various feed items. For desert and hibernating species, metabolic water is their only supply of water; for marine creatures, feed water is their main source of water. Water is essential to all living things. In adult animals, water makes up half to two-thirds of the body mass, whereas in newborn animals, water makes up over 90% of the body mass. Water is a necessary component of practically all bodily secretions. Water is a ubiquitous solvent in the body that helps with cellular biochemical events that include nutrient digestion, absorption, and transportation. Water's aqueous medium facilitates the interaction of various digestive juices and dietary ingredients, improving digestion. It also aids in the excretion of waste products from the animal body, including urine, feces, and sweat. Water's high specific heat allows it to absorb heat produced by various physiological processes, which aids in controlling body temperature. Water also controls body temperature by transferring heat from organs through blood or by evaporating as sweat. Body cells are shaped by water. Water aids in preserving the body's acid-base equilibrium. Water cushions the nervous system and tissue cells, shielding the numerous critical organs from shocks and injury.

### 2.6.1. Water Sources and Losses

Animals obtain water from various sources, including drinking water, moisture in food, and water released during metabolic processes. The relative importance of these sources varies by species, habitat, and diet. For instance, marine creatures meet their water needs from their food, while hibernating species and desert rodents rely on metabolic water to survive. The amount of metabolic water produced also depends on the type of nutrient being metabolized, with fat oxidation producing the most metabolic water. However, in most species, metabolic water accounts for less than 5% to 10% of daily water needs.

Ruminant and non-ruminant animals differ significantly in the water content of their diets. Ruminants consume a variety of forages, ranging from 5% to 7% water content in mature plant products, such as hay, to over 90% in fresh, lush green vegetation. Sheep and other animals heavily rely on the water content in these verdant forages. In contrast, the moisture content in commercial diets for non-ruminants, such as pigs and poultry, ranges from 7% to 10%. More than 75% of the food in canned diets for pets like cats and dogs may consist of moisture.

Water is lost from the body through processes like breathing, lung evaporation, urine, feces, sweat, saliva, and milk in lactating animals. The greatest water loss typically occurs through urine, which helps remove toxic byproducts of metabolism. Certain species, like birds, conserve water by concentrating their urine and excreting it as uric acid rather than urea. Urinary water loss is influenced by diet and environmental conditions; heat stress and excessive water intake can increase urinary volume. Animals consuming high-fiber diets excrete more water in their feces. For example, cows lose 30%–32% of water through feces, whereas sheep lose 13%–24%, with their pellet-shaped feces minimizing water loss.

Sweating is another way animals release heat, and in some species, like horses, it leads to substantial water loss. Animals with poorly developed sweat glands, such as dogs and chickens, rely on increased water consumption and panting to manage heat loss. Providing clean water daily is crucial during periods of heat stress, particularly for animals like chickens, to offset water losses.

### 2.6.2. Water Requirements

The amount of water an animal needs varies depending on several factors, including outside temperature, diet (energy content, fiber content, and salt), physiological status (age, growth, pregnancy, and capacity to store water), level of exercise, and overall health. One important factor influencing water intake is the surrounding temperature and the humidity that accompanies it.

In non-heat-stressed, non-lactating calves, water consumption is measured in units of body weight and ranges from 5% to 6% of body weight per day (or 2.5 kg of water for every kilogram of dry feed consumed), increasing to 12% or more in cases of heat stress. Due to evaporative water loss, water intake rises in response to temperature increases and increased physical activity. At moderate temperatures, there is a strong correlation between feed water content, dietary dry matter intake, and water consumption. Diets high in calories, fat, and protein result in an increase in water consumption due to the production of heat during metabolism, as well as increased metabolic waste and urea excretion in the urine. Water consumption also rises with increased dietary salt content. Fiber-rich diets (bran, dry forages) lead to an increase in water consumption as well. Larger animals have higher water requirements compared to younger animals, as they have greater body sizes. Similarly, animals that conserve water, such as chickens and lambs, require less than cattle. Water consumption increases during pregnancy and lactation. Dairy cattle may need 38–110 L/d, while beef cattle typically require 22–66 L/d (Table 2.11).

Table 2.11. *Animal's Water Requirement: Number of Animals and Global Average Daily Requirements for Service Water and Drinking per Head*

| **Species** | **Herd Size (Million Heads)** | **Service Water (L/Day/Head)** | **Drinking Water (L/Day/Head)** |
|---|---|---|---|
| Buffalo | 200 | 5 | 60 |
| Cattle | 1,460 | 4 | 50 |
| Chicken | 21,900 | 0.05 | 0.4 |
| Goats | 1,010 | 4 | 3 |
| Pigs | 992 | 20 | 10 |
| Sheep | 1,190 | 4 | 5 |

*Source: Wisser, D., Grogan, D. S., Lanzoni, L., Tempio, G., Cinardi, G., Prusevich, A., & Glidden, S. (2024). Water use in livestock agri-food systems and its contribution to local water scarcity: A spatially distributed global analysis. Water, 16(12), 1681. https://doi.org/10.3390/w16121681.*

### 2.6.3. Water Restriction and Toxicity

Animals, both domestic and wild, are impacted by water scarcity. Pure water is necessary for animals to replenish their losses and to sustain the associated physiological processes. Insufficient water may cause feed intake and productivity to decline. Body weight decreases as a result of dehydration, and the effects are exacerbated in warm environments. Loss of electrolytes, a rise in body temperature, and an increase in breathing rate are all signs of dehydration. After chronic water scarcity, animals become extremely agitated and eventually die of prostration. Water intoxication is caused by the kidneys' sluggish tolerance to the high-water load and can happen when a person suddenly consumes significant volumes of water after a brief period of deprivation.

### 2.6.4. Water Quality

Although water is nearly harmless on its own, contaminants such as pesticides, minerals, bacteria, and parasites can compromise water quality. Water quality impacts animal health, productivity, and consumption. Salts, pathogens, algae, pesticides, and other substances can contaminate water sources and affect the taste of food. Mineral salts include calcium, magnesium, sodium, potassium, sulfate, and bicarbonate. Other harmful elements in water include hydrocarbons, iron salts, and nitrate. In regions with intensive agriculture, nitrate contamination is common. Toxicity leading to anoxic death may occur at concentrations above 1,500 ppm. Iron salts in groundwater can lead to rust accumulation on pipes and bacterial contamination from iron-utilizing bacteria.

Toxic pesticides, such as malathion and organophosphates, may enter water systems. Certain cyanogenic blue-green algae found in lakes can produce toxic chemicals. Blue-green algae toxicosis has been known to cause vomiting, foaming at the mouth, muscle tremors, liver damage, and even death in livestock. Water quality can also influence polioencephalomalacia, a non-infectious brain disease that affects cattle in feedlots. Affected animals often exhibit blindness, head swelling, a stargazing posture, aimless wandering, and disorientation. Sulfate-rich water appears to contribute to polioencephalomalacia through complex interactions with other minerals and B vitamins.

Domestic animals can generally tolerate total dissolved solid (TDS) levels of 15,000 to 17,000 mg/L. Water containing less than 1,000 mg/L of total soluble salts is safe for all livestock. However, higher concentrations (>5,000–7,000 mg/L) may increase mortality in poultry and cause mild diarrhea, while other livestock may tolerate it (Table 2.12).

**Table 2.12.** *Water Withdrawal for All Crops from WBM and Percentage Used for Feed After the Allocation*

| | Direct Water Withdrawal | | Feed Water Withdrawal | | Total |
|---|---|---|---|---|---|
| **FAO Region** | **[$km^3$]** | **Percent of Total [%]** | **[$km^3$]** | **Percent of Total [%]** | **[$km^3$]** |
| Australia and New Zealand | 1.18 | 15 | 6.74 | 85 | 7.92 |

| | | | | | |
|---|---|---|---|---|---|
| Caribbean | 0.259 | 20 | 1.02 | 80 | 1.27 |
| Central America | 1.41 | 10 | 12.2 | 90 | 13.6 |
| Central Asia | 0.683 | 2.7 | 24.4 | 97 | 25.1 |
| Eastern Africa | 2.88 | 51 | 2.72 | 49 | 5.6 |
| Eastern Asia | 9.17 | 5.4 | 161 | 95 | 170 |
| Eastern Europe | 1.77 | 7.4 | 22.1 | 93 | 23.9 |
| Middle Africa | 0.886 | 83 | 0.178 | 17 | 1.06 |
| Northern Africa | 1.77 | 13 | 11.5 | 87 | 13.3 |
| Northern America | 4.16 | 6.3 | 61.6 | 94 | 65.8 |
| Northern Europe | 0.968 | 82 | 0.218 | 18 | 1.19 |
| South America | 8.89 | 43 | 11.7 | 57 | 20.6 |
| South-Eastern Asia | 2.51 | 7.9 | 29.1 | 92 | 31.6 |
| Southern Africa | 0.519 | 27 | 1.38 | 73 | 1.89 |
| Southern Asia | 10.3 | 7.5 | 127 | 92 | 137 |
| Southern Europe | 1.19 | 6.5 | 17 | 93 | 18.2 |
| Western Africa | 1.99 | 63 | 1.17 | 37 | 3.16 |
| Western Asia | 1.02 | 5.5 | 17.4 | 94 | 18.4 |
| Western Europe | 1.66 | 25 | 5.11 | 75 | 6.77 |
| **Total** | **53.2** | **9.4** | **513** | **91** | **566.2** |

*Note(s): Regions with a contribution to total water withdrawal of less than 0% were excluded from the table, namely Melanesia (0.034 km$^3$) and Micronesia (0.0000003 km$^3$).*

*Source: Wisser, D., Grogan, D. S., Lanzoni, L., Tempio, G., Cinardi, G., Prusevich, A., & Glidden, S. (2024). Water use in livestock agri-food systems and its contribution to local water scarcity: A spatially distributed global analysis. Water, 16(12), 1681. https://doi.org/10.3390/w16121681.*

## 2.7. FEED ADDITIVES

### 2.7.1. Enzymes

In order to lessen the antinutritional effects of the feed, exogenous enzymes are frequently given to the diet of monogastric animals. Plant-based byproducts and cereal grains make up the majority of the ration. These items contain some nutrients that are in a "bound" form, which means that monogastric animals lack the enzymes necessary to break down and release the nutrients so the animal can use them. The two main antinutrients found in feed are non-starch polysaccharides (NSP) and phytate phosphorous. Phytic acid, also known as myo-inositol hexakisphosphate, is the main phosphate storage molecule found in seeds and is found in cereal grains. Usually, between 50% and 80% of the total phosphate in plant seeds comes from phytic acid. Phytate is the salt form of phytic acid, and nearly all phytic acid exists as a mixed salt (phytin). Due to the absence of endogenous enzymes, phytate P is poorly accessible to monogastric animals and can decrease the digestibility of other nutrients, which can have an impact on growth and performance. The most often utilized exogenous enzyme in animal feed intended for monogastric consumers is microbial phytase. The gradual extraction of phosphate from phytic acid or its salt, phytate, is catalyzed by the enzyme phytose, also known as myoinositol hexakisphosphate phosphohydrolase. Phytase can lessen the detrimental effects of inorganic P excretion to the environment and enhance the digestibility of phosphorous (P), calcium, amino acids, and energy. It can also lessen the antinutritional effect of phytate. Because phytate solubility differs among feedstuffs, phytase effectiveness also changes. Overall, the animal (gut pH, retention period) and feed (enzyme source, solubility, and particle size) aspects influence the effectiveness of exogenous phytase enzymes. NSPs are considered feed ingredients that are either poorly digested or have antinutritive qualities. They make up a large portion of fiber, which is composed of lignin and NSP together. Water-soluble and insoluble substances like cellulose, lignin, arabinoxylan, glucan, and oligo- and polysaccharides with various sugars like arabinose, galactose, rhamnose, and mannose are examples of NSPs. In rye and wheat, arabinoxylans constitute the predominant NSP, whereas barley and oats have a substantial (3–4%) content of mixed-linked β-glucan. Reduced development and performance in monogastric animals are linked to high amounts of NSP in their diet. In particular, high-NSP diets promote the formation of pathogenic bacteria because they raise the viscosity, reduce the mixing and digestibility of the digesta, and slow down the transit of the digesta through the digestive system. Enzymes called carbohydrases are specific to sugars and complex carbohydrates (NSPs). It has been found that a number of enzymes, such as cellulase, ß-glucanase, and xylanase, increase the nutritional value of grain leftovers for pigs and fowl. However, the solubility of the NSP and the complex nature of the carbohydrate determine how effective the enzymes are. The development of highly effective enzymes that enable monogastric animals to use both soluble and insoluble NSPs as an energy source will be a future issue for the feed industry.

## 2.7.2. Pro/Prebiotics

Probiotics and prebiotics are terms used to describe both live and nonliving species and/or substances that support a balanced microbial population in the intestine. Encouraging a healthy gut microbiota is crucial, particularly as animal feed starts to exclude antibiotic growth promoters (AGPs). Through a variety of processes, probiotics aid in the prevention and management of gastrointestinal infections and/or enhance the well-being and productivity of industrial animals. Animal performance is frequently linked to healthy bacteria populations in the gastrointestinal (GI) tract, which may indicate better immunity and more effective digesting. Live microorganisms, both bacterial and nonbacterial, are known as probiotics. Most of the microorganisms employed are bacterial in origin, with the exception of a few yeast and fungal probiotics. Several species of Lactobacillus, Bifidobacterium, Bacillus, and Enterococcus are examples of bacterial probiotics. Probiotics that are not bacterial (yeast or fungal) include Saccharomyces, *Candida pintolopesii*, and *Aspergillus oryzae*. It has been demonstrated that taking probiotic supplements increases the number of good microorganisms like Bifidobacteria and Lactobacilli. The development of pathogenic germs is subsequently prevented by these healthy bacterial colonies by the production of compounds (such as bacteriocins and/or organic acids) and by competitive exclusion. Because probiotics inhibit intestinal colonization and the spread of other enteric pathogens, they can be used as a substitute for antibiotic feed additives in managing and lowering pathogen load. Probiotics have been demonstrated to be useful in lowering pig morbidity and mortality as well as postweaning diarrhea in piglets. Prebiotics are indigestible functional nutrients that specifically promote the growth of beneficial bacteria in the host's digestive system (think of them as food for the good guys). The most widely utilized prebiotics include lactulose, mannan-oligosaccharide (MOS), inulin, and fructooligosaccharides (FOS). Prebiotics are categorized as disaccharides, oligosaccharides, and polysaccharides.

## 2.7.3. Antioxidants

The need for antioxidant protection has been made abundantly clear by the introduction of meals with high levels of animal and vegetable fats. Feeds and feedstuffs can become oxidized, which can lead to rancidity of fats, the breakdown of vitamins A, D, and E, and the loss of amino acids. This can reduce the amount of energy in the diet. Animal diets are supplemented with antioxidants to enhance feed quality and prolong shelf life. Artificial and natural compounds are the two categories of antioxidants. These products offer protection against reactive oxygen species by having the capacity to stop or prevent the generation of free radicals. Antioxidants reduce oil rancidity and improve feed quality and shelf life through these mechanisms. Tocopherols are one type of natural antioxidant; other aromatic plant extracts (like rosemary extract) and essential oils are next in line. RRR-α-tocopherol is vitamin E in its natural form. Synthetic vitamin E, also known as all-rac-tocopherol or all-racemic-α tocopherol, is composed of various stereoisomers. One milligram of allrac-α-tocopherol acetate is equivalent to one IU of vitamin E. Commercial forms of vitamin E in the feed business include RRR-α-tocopherol acetate and all-rac-α-tocopherol acetate. Butylhydroxytoluene (BHT), butylhydroxyanisole (BHA), and ethoxyquin (generic name for 1,2-dihydro-6-ethoxy-2,2,4-trimethylquinoline) are further synthetic chemicals possessing antioxidant qualities. But ethoxyquin has been shown to be the most effective, with BHT and BHA coming in second and third.

## 2.7.4. Antibiotic Growth Promoters

When hens were fed feed containing tetracycline fermentation byproducts in the 1940s, the growth-promoting properties of antibiotics were found. Since then, a variety of antibiotics are used as growth promoters on multiple species, extending their original application. Antibiotic growth promoters are compounds added to animal feed at low subtherapeutic dosages to improve animal performance and reduce illness. Because of advancements in feed conversion to animal-derived foods and rising consumer demand, the usage of AGPs has expanded along with the intensity of livestock production. Numerous natural, synthetic, or semisynthetic veterinary antibiotics possessing antibacterial properties are employed as feed additives for preventive disease management or as growth boosters in animal production. These items have been shown to reduce foodborne infections and improve animal health and food output. Animal feeds are supplemented with substances other than antibiotics. In intensive meat production, anabolic steroids and β-agonists are commonly utilized to enhance nutritional availability and lean meat deposition in food animals. In the dairy sector, recombinant bovine somatotropin (rBST) has been utilized to increase milk output. A class of drugs known as β-agonists, which includes citamaterol, acts on adrenergic receptors to cause anabolic effects in animals. Anabolic steroids are employed in animal production systems to boost growth rate, feed efficiency, and carcass leanness. They also improve slim meat deposition, muscle protein accretion, and fat stores' metabolism. The European Union has outlawed anabolic feed additives; in the US, however, melengestrol acetate for feedlot heifers and ractopamine for swine are permitted as feed additives. Overuse of antibiotics, especially when administered at subtherapeutic dosages to stimulate growth, can encourage the emergence of antibiotic-resistant bacteria, endangering the health of both people and animals.

Antibiotics and anabolic growth stimulants, which can be especially harmful to aquatic creatures, can also find their way into the environment through animal excrement and wastewater. Antibiotic producers in the United States have voluntarily removed AGPs from the market for nontherapeutic use. Many nations control its usage for promoting animal development, although they do not outright forbid it.

## 2.7.5. Others

Depending on the species, feed type, and market, the feed may also contain a number of additional minor components in addition to the major additives already mentioned. They include mycotoxin binders, which bind to the mycotoxins and prevent absorption, coccidiostats, which prevent coccidiosis (a protozoan infection in poultry), and pellet binders, which lessen feed dust and improve pellet adhesion. Natural or synthetic pigments, such as carophyll, lutein, and zeaxanthin, are used to enhance the color of egg yolks and grill skin.

Chapter 3

# Feeding Management of Ruminant Livestock

## Contents

## 3.1. OVERVIEW

The livestock industry contributes significantly to the national economy and is expanding at a faster rate. One of the most crucial aspects of animal care is feeding. It is widely acknowledged that when fed at low levels, all animals are unable to fulfill their genetic potential for increased production. Young stock that is underfed experiences stunted growth, delayed maturity, and subpar productivity once they reach the breedable age. Feed expenses might account for as much as 65–75% of the overall cost of production. The producers' incomes are also increased by the high-quality feed. A balanced diet for the animal is one approach to cut these expenses. A balanced diet is one that satisfies the livestock's needs for nutrients according to its age, gender, and physiological stage. Sufficient nutrition is necessary for any animal's health and metabolic processes. Long-term nutritional deficiency would cause productivity loss and condition loss. Inadequate feeds can result in a deficiency of some dietary components, or other elements can trigger the emergence of severe nutritional disorders. Just as harmful as underfeeding can be overfeeding. On the other hand, avoiding dangerous substances from getting into the food chain and harming people's health also depends heavily on the safety and quality of animal feedstuffs. The qualities and content of animal products such as meat, milk, eggs, and so forth for human consumption, as well as the welfare of the animals, are influenced by feedstuffs and additives, diet formulation, and, in certain situations, diet distribution.

Nonetheless, the main cause of decreased milk production in dairy cows belonging to distinct breeds is insufficient nutrition. According to multiple sources, there is a shortage of commercial concentrates, green fodder, and dry fodder, which prevents animals from receiving enough nutrition to express their genetic potential for producing milk. The lack of or insufficient usage of commercial concentrate feeds, the low quality and nutritional imbalance in many native pastures and agricultural leftovers, and the shortage and inadequate amount and quality of feed supplies are the key obstacles to livestock development in many developing nations.

## 3.2. IMPORTANCE OF NUTRIENTS

The nutritional state of the animals and the methods used for management determine how efficiently animals are produced. In terms of animal health and productivity, nutrition is crucial because:

- It highlights the animal's genetic potential. For instance, a cow cannot produce 25 liters of milk per day if it is underfed, even though its genetic composition indicates that it should be able to produce that much milk.

- Because the cost of feeding animals makes up 70–75% of the entire cost of producing animals, it makes animal production economical.
- By using non-traditional feeds for animal feeding, it reduces the competition between humans and animals for food.
- It modifies feed components to maximize nutrient utilization.

Factors affecting the chemical composition of plants include:

1. **Plant Factor:** Because distinct genetic material makes up different types of the same species of forage, its chemical composition varies.
2. **Agro-Climatic Condition:** The chemical makeup of plants is influenced by temperature and humidity levels in the atmosphere.
3. **Cultivation Practices:** Plant growth rate and chemical composition are influenced by various factors such as seed rate, seed treatment, timing and technique of sowing, irrigation, manure and fertilizer, and control of weeds and diseases.
4. **Stage of Growth:** While the amount of dry matter and crude fiber increases as the plant ages, the amount of soluble ash and crude protein is higher right before flowering and decreases at the seed development stage. When the plant reaches maturity, ether extract gradually decreases.
5. **Processing and Preservation Practices:** Plant materials can undergo changes in their composition, nutrient content, size, and shape due to different processing techniques.

## 3.3. RUMINANT DIGESTIVE SYSTEM AND DIGESTION

Mammals that consume plants, such as sheep, goats, deer cattle, buffaloes, bison, giraffes, yaks, and so on, are referred to as ruminants. These cud-chewing mammals digest their food in their four-chambered stomachs. Through a process known as rumination, some species have adapted to eat plant materials. Prior to the primary digestive process, they chew, regurgitate, and ferment their food through the action of rumination. Human digestion is entirely different from that of ruminants. The digestive process of food generally starts in the mouth and moves from there to the food pipe, stomach, and intestines. This procedure is entirely different in ruminants. Rats consume food rapidly. They cannot, however, directly digest grasses, foliage, or other plant material because their systems do not manufacture the enzymes necessary to break down cellulose. In a stomach with multiple chambers, digestion happens one chamber at a time. In the rumen, the first chamber of a ruminant's stomach, which has four chambers, consumed plant material is kept until it softens. They subsequently chew and regurgitate this substance, known as cud, in an effort to further break down its tough-to-digest cellulose component. The chewed cud travels straight to the stomach's other chambers, the reticulum, omasum, and abomasum, where it is further broken down by a variety of vital microbes. Although the stomach of a pseudo-ruminant is separated into three chambers, the digestive process is quite similar, involving both chewing of the cud and regurgitation. The digestive system of ruminants is distinct because of its several stomach chambers, which comprise the reticulum, omasum, abomasum, and rumen. Through the use of a sophisticated system, they can transform plant-based diets that are indigestible to many other animals into beneficial nutrients.

Numerous morphological adaptations are present in ruminant animals to aid in the processing and digestion of the chemical compounds found in plant cell walls. The animal takes in food through its mouth, which is also where the first anatomical adaptations for grasping, cutting, and grinding food items are found. The size and form of ruminants' mouths have a direct impact on how much food they eat and which foods they choose to consume. Because of their highly flexible lips, large gripping tongues, more projecting incisors, and narrow incisor arcades, browsers are better able to identify and hold small food items within complicated plant structures like leaves and shoots. Cattle are a good example of a species that mostly feed on grasses; they have broader muzzles and incisor arcades to maximize bite size inside homogenous grass swards. A variety of diets with varying mass, abrasiveness, and mechanical resistance to comminution have also been adapted to molars. In comparison to their body size, grazers have larger molar occlusal surfaces to grind tough and coarse foods, and they have hypsodonts (teeth with a high crown), which strengthen and withstand abrasions on the molars. In contrast, browsers have narrower and smaller post-canine teeth than grazers, but they also have more noticeable dental crests that are designed to break through plant cell walls. While some feeding patterns are connected with larger-bodied species—grazers, for instance, have larger bodies than many browsers or mixed feeders—certain functional adaptations hold true even when body mass is taken into account. Only body size and hypsodonty, however, continued to be significant adaptations to feeding patterns when body mass and phylogeny were taken into account in the studies. In comparison to mixed feeder and grazer species, browsers are smaller and possess shorter and less bulky molars (Pérez-Barbería et al., 2004). The salivary glands have undergone a noteworthy modification in response to the feeding pattern. Browse includes secondary chemicals that, by binding to proteins and reducing enzymatic digestion, can significantly affect apparent protein digestibility. These compounds include phenolics, alkaloids, and terpenoids. To minimize the detrimental impact on plant fiber digestibility, browsers possess highly developed mandibular and parotid salivary glands that release chemicals to counteract the secondary compounds' ability to bind to proteins (Figure 3.1).

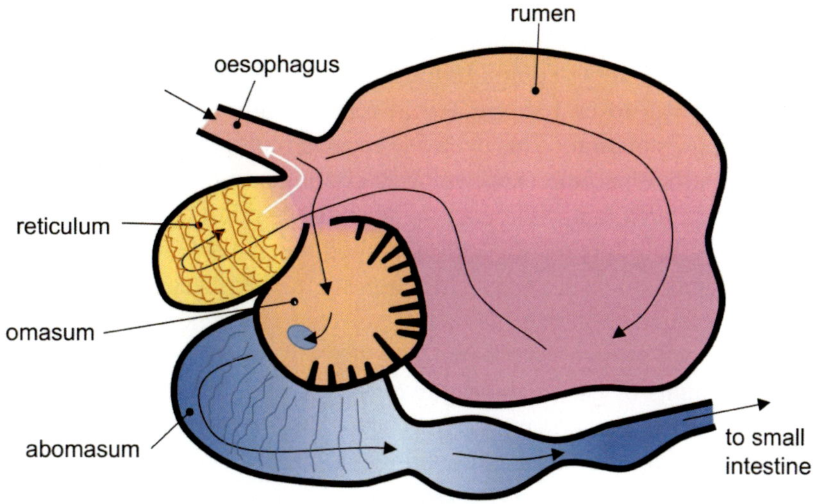


Figure 3.1. *Basic morphology of the ruminant stomach.*

*Source: García and Pérez (2020).*

The stomach, which is connected to the mouth by a wide esophagus, is the primary anatomical feature that distinguishes ruminants. Within the ruminant digestive system, the stomach is a chamber that provides an ideal habitat for the growth of a diverse variety of bacteria, protozoa, and fungi. The ruminant depends on bacterial symbiosis to metabolize cell walls since vertebrates lack the enzymes necessary to break down the structural carbohydrates that makeup plant cell walls. Through enzymatic digestion in the abomasum and small intestine, the ruminant itself may readily metabolize other highly digestible substances like carbohydrates, proteins, and lipids. Although microorganisms colonize the stomach of all vertebrate species, ruminants have developed large-volume, multi-compartmentalized stomachs that allow a wide variety of microorganisms to flourish in such quantities that the ruminant is not only able to use the byproducts of the metabolic activity of the microorganisms but also depends entirely on them for its survival.

The rumen is the largest chamber, with an internal papilla covering that increases its absorptive surface area by up to 30 times compared to a non-papillated one. It is distinguished by a thick wall of keratinized epithelium and muscle pillars. Plant material is ingested into the rumen through the esophagus after being cut, chewed, and combined with saliva. To keep the pH of the rumen constant, ruminants generate enormous amounts of saliva that are rich in bicarbonate (up to 150 l/day in cattle). The rumen offers a consistent, anaerobic environment (pH = 6.0–6.4; temperature = 37.5–42°C) that is favorable to fermentation. The rumen walls are made of muscle, which facilitates good peristaltic movement and mixing of the plant material that has been chewed. To promote comminution, this motility also makes it easier for large food particles to be regurgitated into the mouth as a bolus that can be chewed again (rumination). This cycle is repeated multiple times until the meal is reduced to a fine particle size, which is contingent upon the molars' wear, occlusal surface area, and efficiency of chewing. Additionally, rumination is known as merycism in camelids and macropod marsupials. Within the rumen, gases, solids, and liquids are arranged according to their specific gravities. The dense, finely ground plant material and liquids are on the bottom layer; on this layer, the less well-ground, coarse material floats like a raft and gases rise to the top. Huge amounts of gas, including hydrogen, carbon dioxide, and methane, are created during the fermentation process. The reticulum, a compartment with a honeycomb reticulated epithelium that generates tiny pools to aid in the retention of particles of varying densities and sizes, receives the dense and finer material. The reticulo-omasal aperture allows the reticulum to open into the omasum. The omasum is distinguished by consecutive longitudinal walls of epithelium that hold onto the larger and less digested food particles while further facilitating the passage of tiny food particles and the attached ruminal microorganisms that are breaking down cell walls. Additionally, the omasum divides the very acidic abomasum environment—where water, soluble particles, and microbial activity products are absorbed—from the rumen-reticulum environment. The fermentation products either pass down to the small intestine for additional digestion and absorption, or they are absorbed through the rumen walls and into the bloodstream.

The small intestine, where the majority of enzymatic digestion and nutrient absorption that has evaded microbial action occurs, is connected to the abomasum. There are three distinct zones: the distal ileum, the intermediate jejunum, and the proximal duodenum. The largest small intestines in mammals are found in herbivorous animals, especially ruminants, and are followed by granivores and frugivores. Short intestines are found in animals like carnivores and insectivores that consume highly nutritious and easily digested foods. Within a species, seasonal variations in diet also affect the length of the small intestine. Plant components that are tough to digest can be more easily absorbed and digested because larger small intestines allow food to travel through the digestive tube more slowly. In browser and grazer ruminants, the length of the small and large intestines can add up to 12 and 30 times, respectively, to the body length. The liver, pancreas, intestinal mucosa, and duodenal glands are the four organs that provide the small intestine with digestive secretions. These secretions allow a range of enzymes, bile salts, and buffering solutions to be combined for the control of pH and chemical reactions. Certain enzymes are generated as proenzymes, which require a sequence of intricate interactions in the intestine to transform into active forms. Neonatal ruminants' small intestines have developed adaptations that, when combined with the mother's colostrum's enzyme inhibitors, allow intact colostral immunoglobulin molecules to be absorbed through pinocytosis and pass on passive immunity to the young. Certain saccharides are absent in ruminants, pseudo-ruminants, and macropod marsupials because the forestomach ferments a large number of disaccharides into volatile fatty acids. Similar to this, bacteria hydrolyze the fatty acids in dietary glycerides and ferment the glycerol fraction, which limits the amount of fat that can be broken down by pancreatic lipase and bile salts in the forestomach (Robbins, 1993).

Diverticula called caeca are found where the small and big intestines converge. There is only one caecum, a sizable and useful digestive organ, in ruminants. At the ileo-caecal junction, the caecum opens up to the large intestine and finishes at the rectum. Similar to the small intestine, a long, large intestine slows down food passage, enhancing plant material that is resistant to digestion's absorption and digestion. Up to 78% of the small intestine's length is made up of the big intestine. The caecum and large intestine are the sites of further fermentation of fiber, endogenous matter, absorption of water and water-soluble minerals, amino acids, volatile fatty acids, ammonia, electrolytes, and bacterial production of vitamins. Simultaneously, the large intestine and caecum gather leftover material from the digestive process and get it ready for feces production (Robbins, 1993) (Table 3.1).

Table 3.1. *Physicochemical Properties of the Rumen*

| **Parameter** | | **References** |
|---|---|---|
| Temperature | 39°C (optimal), vary in between 38–41°C | (2, 8) |
| pH | ~6.5 (buffered in the range of 5.5–7.0) | |
| Dry content | Maintained constant around 10–20% | |
| Osmolality | 250–400 mOsmol/Kg (increases with the feed intake) | |
| Redox potential | Lies within the range of −150 to −350 mv | |
| Gaseous composition | $CO_2$ (65%), $CH_4$ (27%) are the major gases produced by fermentation. $N_2$ (7%), $O_2$ (0.6%) $H_2$ (0.2%) are present in traces. | |

*Source: Xu, Q., Qiao, Q., Gao, Y., Hou, J., Hu, M., Du, Y., Zhao, K., & Li, X. (2021). Gut microbiota and their role in health and metabolic disease of dairy cow. Frontiers in Nutrition, 8, 701511. https://doi.org/10.3389/fnut.2021.701511.*

## 3.4. FEED STUFF

Food supplied for cattle and other livestock is referred to as feed stuff. It also functions as any one of the nutrient components of an animal feed. Feedstuffs are consumable materials that animals can digest, absorb, and use after consuming them. Feedstuffs are made up of dry materials and water. Feedstuffs' water (moisture) content varies greatly and can be as high as 900 g/kg (in certain root crops) or as low as 60 g/kg (in concentrates). It is usually suggested that the feedstuff composition be given on a dry matter basis because of this significant fluctuation. From this vantage point, feedstock nutrient levels may be efficiently compared. Animal production performance can be significantly impacted by the nutritional components of a diet. The primary nutritional components of a feedstock are measured by its feed value (Figure 3.2).

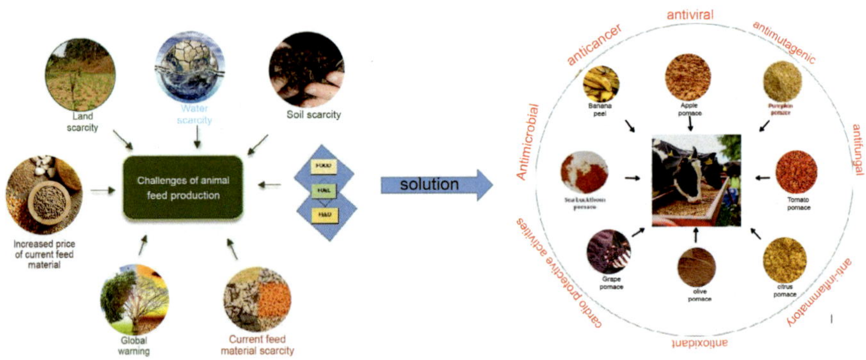


Figure 3.2. *The composition of feedstuffs.*

*Source: Malenica, D., Kass, M., & Bhat, R. (2023). Sustainable management and valorization of agri-food industrial wastes and by-products as animal feed: For ruminants, non-ruminants, and as poultry feed. Sustainability, 15(1), 117. https://doi.org/10.3390/su15010117.*

The concentration of energy (carbs, fats, proteins, and their digestibility), protein (including NPN and aspects of degradability), vitamins, and minerals in the dry matter; special aspects (like keeping quality, availability, handling, taste, toxins, influence on sensory quality of meat, milk, or egg, etc.); physical aspects; and price are the main factors that determine the feed value of any feedstuff for livestock. However, a number of vitamin and mineral shortages, such as those in calcium, magnesium, phosphorus, copper, cobalt, vitamins A or D, and so on, can severely limit production. Furthermore, reduced productivity and even death might result from feedstuffs containing excessive amounts of specific chemicals. Feeds are categorized based on the number of particular nutrients they provide. There are two primary categories of feedstuffs: concentrate and pasture and roughages. Furthermore, feeds can be categorized even further, as indicated in Figure 3.3.

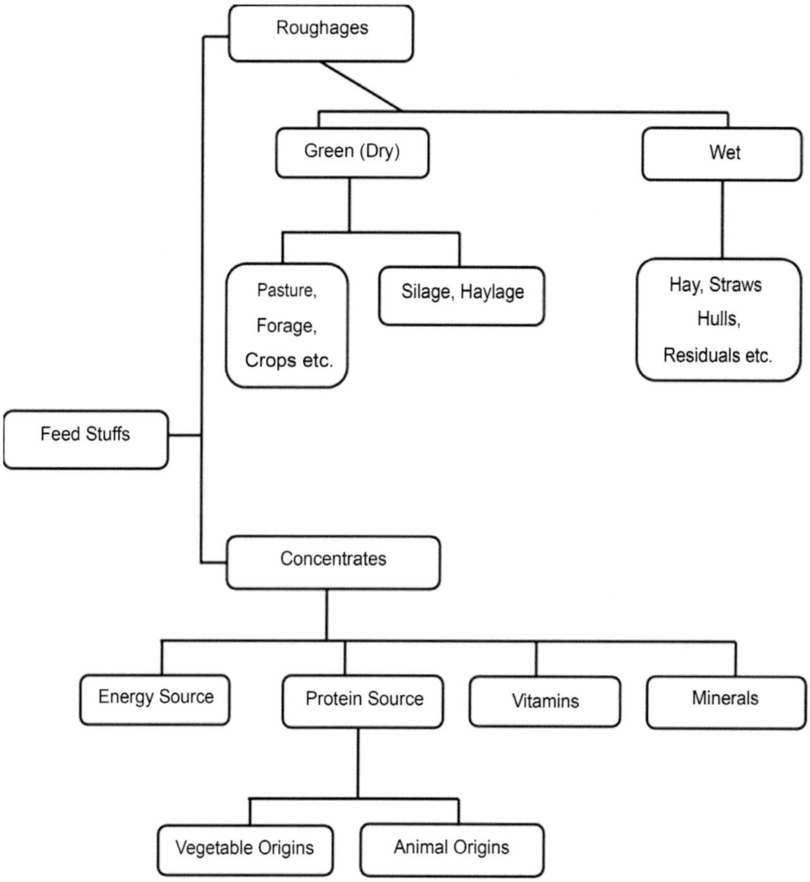


Figure 3.3. *Feed stuff classification.*

*Source: Yucel and Taskin (2018).*

Roughages are dense feeds with a significant proportion of indigestible substances, containing more than 18% crude fiber. They can be divided into two main categories based on moisture content: dry and wet. Dry roughages, which contain only 10–15% moisture, include hays, straws, hulls, and crop residues like seed coats, pods, and bran. Wet roughages, with more than 75% moisture, include pasture, range plants and forages fed green, fodder crops, grasses, legumes, tree leaves, and silage/haylage. Ensiled forages, a type of silage or haylage, are preserved through the ensiling process, which involves storing plant materials in anaerobic conditions—a common feed storage technique.

During controlled fermentation, acids are produced that destroy bacteria, fungi, and other harmful organisms. However, this fermentation process reduces the nutrient content of the material.

Energy-rich diets contain less than 18% crude fiber and less than 20% protein. Protein digestibility varies between 50% and 80%, with most proteins being of low quality. These energy sources are fed primarily to monogastric animals as their main energy supply and to ruminants and cecal fermenters to boost energy density in their diets. Examples of energy sources include cereal grains (such as corn, wheat, barley, oats, rye, sorghum, and triticale), other grains like buckwheat, grain milling by-products like wheat bran and corn gluten meal, roots, and tubers like cassava and potatoes, food processing by-products like molasses, bakery waste, citrus pulp, distillers' and brewers' by-products, and industrial by-products like wood molasses, fats, and oils.

Protein supplements contain 20% or more protein and, depending on the source, significant levels of energy from either plants or animals. Examples of protein sources include oilseed meals (soybean, cottonseed, rapeseed, canola, linseed, peanut, safflower, sunflower), grain legumes (beans, peas, lupines), single-cell proteins, synthetic amino acids, non-protein nitrogen sources (urea, biuret), by-pass proteins (e.g., ruminant corn gluten meal), and animal proteins (meat, fish, tankage, feather, bone, dried milk, or products like whey and poultry by-products).

Concentrated sources of vitamins and minerals may be required to balance a ration, depending on the other foods used. Examples of vitamin supplements include ensiled yeast, fish oil, wheat germ oil, liver meal, and purified forms of specific vitamins (A, D, E, K, C, and B vitamins, including thiamine, riboflavin, niacin, pantothenic acid, biotin, vitamin B6, vitamin B12, and folate). Mineral supplements include salt (often trace mineralized), oyster shell, bone meal, calcium carbonate, limestone, and pure forms of specific minerals. Major elements include sodium (Na), calcium (Ca), phosphorus (P), chloride (Cl), potassium (K), sulfur (S), magnesium (Mg), iodine (I), manganese (Mn), fluoride (F), cobalt (Co), boron (B), zinc (Zn), iron (Fe), copper (Cu), and molybdenum (Mo) as trace elements. Since their rumens are functional, ruminant animals (cattle, buffalo, sheep, and goats) do not require dietary supplements of fat-soluble vitamin K or water-soluble vitamins. Animals exposed to sunlight can produce vitamin D, while those consuming sufficient green forage do not need vitamin A or E supplements. Vitamin and mineral supplements are available under various brand names and can be added to feed or administered through electrified supplements or mixed with drinking water. These supplements need to be fed consistently for 10 days or more. Feed additives are used to promote growth, productivity, and to prevent deficiency disorders.

Although feed additives aren't really nutrients, their minute amounts in feed help in feed intake, digestion, and utilization. Ultimately, they raise the feed's nutritional value, which boosts animal development, productivity, and feed efficiency. A component, or combination of ingredients, added to the basic feed mixture that must be handled and combined with caution, is called an additive. Feed additives are used to lower production costs, safeguard animal health, boost feed conversion, and enhance the quantity and quality of animal products in terms of hygienic standards and quality. It is crucial that additives can be identified in feeds and finished goods to regulate the impact of these compounds on human health. Animal agriculture has seen significant change in recent years. In particular, the European Union has changed feed additives used in the feed sector, taking environmental, human, and animal health concerns into consideration. The trend towards healthier eating habits and a return to natural methods of raising animals has sparked debates about feed additives. At the same time, for instance, the use of substitute feed additives has gained popularity due to issues brought about by the overuse of antibiotics. Among the feed additive categories are:

1. **Technological Additives:** Anything that is added to food in order to advance technology. This includes substances for controlling radionuclide contamination and reducing mycotoxin contamination of feed. It also includes preservatives, antioxidants, emulsifiers, stabilizers, thickeners, gelling agents, binders, anticaking agents, acidity regulators, silage additives, and denaturants.
2. **Sensory Additives:** Any ingredient, such as coloring and flavoring compounds, that when added to feed either enhances or modifies the organoleptic qualities of the feed or the food's appearance.
3. **Nutritional Additives:** These include pro- and vitamin-rich substances, trace element compounds, amino acid derivatives and their salts, urea, and its compounds, and chemically defined substances with comparable effects.
4. **Zootechnical Additives:** Whatever addition that has a positive impact on the environment or the performance of healthy animals. Digestibility enhancers are compounds that, when given to animals, operate on specific feed ingredients to boost the diet's digestibility. Microorganisms or other chemically specified compounds that are fed to animals and positively impact the gut flora are known as gut flora stabilizers. Environmentally beneficial substances and other zootechnical additions.
5. **Coccidiostats and Histomonostats:** These compounds, which have one or more functional groups, are meant to inhibit or kill protozoa.

The preserved substance known as silage is created when crops are carefully fermented in anaerobic environments inside a silo. The method used to make silage is called ensilage. Making silage is mostly done to conserve succulent fodders for use in times of scarcity. Making silage entails allowing natural fermentation to occur in an anaerobic environment while taking precautions to prevent the growth of unwanted microbes. A crop best suited for silage production has a soft, flexible stem. When 50% of the crop is at the ear emergence stage, the crop should be harvested because it will be highly biomass-producing and nutritious at this point. Crops with high moisture content are allowed to wilt for a few hours, or until the moisture content drops to 60%, in order to minimize wastewater loss. The ability to maintain an anaerobic environment in the silo is essential to the success of silage. An anaerobic atmosphere keeps nutrients in crops from oxidizing and creates an environment that is favorable for the survival of beneficial organisms and the production of lactic acid. Therefore, compressing the ensiling materials is crucial to preventing the formation of air pockets in the silo. You can get better compression by chaffing the crop. As silos, a variety of container types are employed. For the silo walls to be robust and smooth, they should be cleaned and replastered. When added to forage at a rate of 2% by weight, molasses provides easily accessed carbohydrates that are essential for lactobacillus to produce more lactic acid. Additional molasses improves the nutritional value and palatability of silage. Quick silo filling is ideal for anaerobic conditions. It is not advisable to make silage on wet days. Since it eliminates air pockets to encourage anaerobic fermentation, compaction is an essential step in the making of silage. As soon as the silo is filled, it should be sealed to maintain the anaerobic state and to keep outside air and rain out. Filling the silo pit to a dome shape and covering it with insulators, such as tarpaulin sheets or mud plaster, is a good idea. Filling in the shape of a dome will let rainwater run-off and stop seepage. Four weeks will pass before the silage is ready. Every day once the silo is opened, the silage needs to be removed to feed the animals. Haylages are low-moisture silages that share traits with both silage and hay. It is composed of 45–55% moisture content grass and/or legumes. The silos should be as airtight as possible to utilize up the oxygen and to capture and retain the $CO_2$ that is created inside the silo. This state will stop the feed from oxidizing, molding, heating, and other spoiling processes. Haylage is a tasty, nutritious meal that smells good. Haylage can be created from forage that has partially dried. Fine cutting, proper packing, and total sealing against air entry into the silo are among the drawbacks of haylages, which are more significant than those of silage. Overheating poses a greater risk of reducing the digestion of protein.

Feeding standards merely specify the amount of nutrients an animal needs to stay healthy and productive. The production (milk, meat, egg, wool), product composition (fat content of milk), and physiological state (growth, fetal development) of animals are taken into consideration while setting feeding requirements. Depending on an animal's breed, age, and intended use, different feeding guidelines may apply. The overall nutritional needs of animals are expressed in modern feeding standards as feed units such as digestible protein, calcium, phosphorus, carotene, and so on. There are also defined guidelines for the consumption of trace minerals, amino acids, and vitamins. Feeding guidelines are subject to periodic revision in response to evolving technology and production objectives. Animal feed formulation is done in accordance with specified feeding guidelines. Feeding guidelines have a long history. As early as 1810, the idea of feeding standards was created in Germany by Greuven, who based his standards on proximate analysis of feedstuffs, and Wolf and Lehmann, who based their standards on digestible nutrients. Subsequently, Morrison standards were developed in the US, periodically amended, and briefly gained popularity. NRC standards, developed by the US National Research Council's Committee on Animal Nutrition, are the current feeding guidelines. They were regarded as the most reliable source of information regarding the dietary requirements of agricultural animals. Subsequently, Sen and Ray created Indian bovine feeding standards, which ICAR has published. The Sen and Ray standards were prepared with reference to the Morrison criteria.

## 3.5. MANAGING DIET AND NUTRITION

In order to guarantee that the feed fulfills the requirements of ruminants during a variety of life phases and environments, effective control of diet and nutrition is crucial. You will find tips in this area on how to create and maintain diets that will optimize productivity and health and save waste and expenses. Several crucial processes are involved in feeding your ruminants the proper diet:

- **Assess Nutritional Needs:** based on the animal's health, desired level of productivity, and life stage (growth, maintenance, lactation, and reproduction).
- **Choose Suitable Feeds:** Choose feeds that meet the specified nutritional needs while taking availability, cost, and palatability into account (as stated in the preceding section).
- **Calculate Rations:** Ascertain the proper proportions of every feed type to provide the requirements for protein, energy, vitamins, and minerals while avoiding overfeeding or underfeeding.
- **Balance the Diet:** Make sure the diet is nutritionally balanced to avoid toxicities and deficiencies, which can result from unbalanced feed ratios.

Effective feed management also involves careful budgeting:

- **Estimate Feed Consumption:** Determine the amount of feed needed over a specific time period based on the quantity of animals and their rates of consumption.
- **Plan Purchases:** Plan your feed purchases to take advantage of bulk buying or savings during times of high availability based on estimations of usage.
- **Monitor Costs:** To maximize financial results, monitor feed costs over time and make necessary adjustments to feeding programs.

Regular monitoring is crucial to ensure that nutritional strategies are meeting the intended goals:

- **Body Condition Scoring (BCS):** By using this easy procedure, you may determine whether your animals' nutritional demands are being met by evaluating their body fat stores.
- **Feed Intake Monitoring:** Maintain feed intake logs to spot trends or variations that could point to medical problems or the need for dietary changes.
- **Health and Performance Records:** To assess the efficacy of your dietary regimen, keep an eye on your health, growth, and milk production rates.

Diet and feed strategies should not be static; they need to adapt to changing conditions:

- **Seasonal Changes:** Diets should be modified to account for variations in the quantity and quality of forage available throughout the year, as well as seasonal factors like the summer's heat.
- **Health Changes:** In response to health problems, alter meals to aid in the recuperation of afflicted animals. For instance, feeding an animal in recovery a food item with a higher calorie density.
- **Feedback from Monitoring:** Make constant improvements to your feeding strategy by utilizing the information gathered from body condition scoring, feed intake, and medical records.

Utilizing technology can greatly enhance your ability to manage diets effectively:

- **Feed Management Software:** Programs that use real-time data inputs on herd performance and feed availability to calculate and modify diets.
- **Precision Feeding Systems:** Measuring and mixing feed ingredients accurately with automated equipment to cut down on waste and

guarantee that every animal is fed a diet that suits its demands at the time.

You may improve farm output, boost economic results, and improve the health of your ruminants by carefully controlling their diet and nutrition. Good nutritional management helps your farming operations remain sustainable while simultaneously meeting the physiological needs of your animals.

## 3.6. ADVANCES IN NUTRITIONAL RESEARCH

Keeping up with the most recent findings in ruminant nutrition research will give you knowledge and resources to increase feed efficiency, promote animal health, and lessen environmental impact. The latest developments in nutritional research that have relevance to contemporary ruminant farming methods are discussed in this section. The effects of genetic variables on an animal's dietary requirements and effectiveness in utilizing nutrients are starting to be revealed by research:

- **Genetic Markers:** More focused breeding strategies can be implemented by identifying the genes that affect characteristics like milk composition and feed conversion efficiency.
- **Custom Diets:** By optimizing health and production based on genetic predisposition, an understanding of genetic variations can lead to more individualized nutrition.

Researchers are always looking for different feed additives and components to make ruminant diets better:

- **Alternative Proteins:** Research is being done on sustainable protein sources such as insect meal, algae, and single-cell proteins.
- **Feed Additives:** Feeds are supplemented with substances like probiotics, prebiotics, and enzymes to help with digestion, increase nutritional absorption, and improve gut health in general.

Technological developments in fodder production enhance pastures' nutritional value and support sustainable farming methods.

- **Biofortification:** Creating fodder crops with improved nutrient profiles, like higher mineral content or better amino acid ratios.
- **Silage Additives:** Utilizing additives to improve the nutritional content and maintain the quality of silage is essential for sustaining the well-being and productivity of livestock over the winter.

Precision nutrition techniques:

- **Nutrigenomics:** Investigating the relationship between DNA and nutrition to develop diet plans that perfectly suit each animal's needs.

- **Precision Feeding Systems:** Technologies that maximize nutrient intake and minimize waste by enabling meals to be changed in real-time in response to an animal's urgent demands.

Sustainability practices in nutrition:

- **Feed Conversion Efficiency:** In order to lower feed costs and the environmental impact of ruminant production systems, research aimed at increasing feed conversion ratios is essential.
- **Reducing Methane Emissions:** Given the substantial contribution of methane to climate change, developing nutritional techniques that reduce methane production by ruminants is an important field of research.

Nutritional health monitoring devices:

- **Wearable Technology:** An animal's nutritional state and general well-being can be instantly determined with the use of devices that track health markers like temperature and pH in the rumen.
- **Remote Sensing:** Evaluating pasture quality and biomass using drone or satellite technologies, which enables more precise and flexible grazing management.

You may increase productivity, boost efficiency, and support more environmentally friendly farming methods by incorporating these advancements into your nutritional management programs. The field of ruminant nutrition is always changing, and keeping up with these advancements can have a big impact on your agricultural business.

## 3.7. FEEDING OF LARGE RUMINANTS

Approximately 70% of the total expenses associated with producing beef cattle are related to feed. An efficient feeding program lowers production costs while raising the productivity and profitability of the cattle industry. To cut production costs, pasture management and quantity are crucial. For maintenance, exercise, development, milk production, reproduction, and overall health, beef cattle need nutrients. These needs for cattle can vary depending on their breed, age, sex, and stage of production. Mature and young calves can obtain the nutrients they need for growth and maintenance if they eat enough high-quality pasture that contains a mixture of grasses and legumes. However, a variety of factors, such as the species of grass, grazing management, soil structure and environmental conditions, and geographic location, will affect the quality of pastures. If these feedstuffs are to be used for any other purpose, they should be supplemented with high-quality concentrate, a vitamin-mineral mixture, and feed additives. With the exception of calves, since they are pre-ruminant, beef cattle can obtain all the energy they need for upkeep from high-quality roughages and forages. Production may require additional energy sources. To ensure maximum performance, healthy reproduction, the wellbeing of the cow and calf, and the growth of all cattle classes, cattle must be fed a sufficient ration that contains the essential nutrients. Microorganisms such as bacteria and protozoa use part of the feed protein to make their own microbial proteins. Rumen degradable protein is a measure of the protein supplied to rumen microorganisms (RDP). The metabolizable protein is required for both production and upkeep. Commercial protein supplements frequently include urea and other non-protein nitrogen (NPN) sources to meet at least one-third of the total nitrogen need. All cattle in normal health have tissues that synthesize vitamin C, and the ruminal microbiota synthesizes enough vitamin K and B complex vitamins. With a few notable exceptions, the mineral element needs for dairy and beef cattle are comparable. Cattle have a relatively low salt (NaCl) need. Cattle should have unrestricted access to water. Water consumption is influenced by a variety of factors, including body weight, temperature, growth, reproduction, lactation, digestion, metabolism, and excretion. Restricting water intake also lowers performance. A significant physiological and biochemical task is lactation. A number of characteristics, including species, breed, strain within the breed, age, and lactation stage, influence the amount and content of milk produced. In order to maintain adequate rumen fermentation and prevent issues with acidosis, reduced intake, and low-fat milk, high energy intakes must have a specific amount of roughage in the diet. The nutritional needs of breastfeeding women are different from those of non-lactating women; they include greater calorie requirements, approximately doubled protein, calcium, and phosphorus requirements, but no change in vitamin requirements. Controlling the quantity and caliber of concentrate during breastfeeding is crucial. Under this approach, the animal's nutritional needs should be sufficiently supplied and overweight should never be permitted. In the absence of this, mid- and late-lactation production performance may decline. Feeding should be inexpensive at the same time. Roughage should make up at least 30% of the entire diet, especially in the early stages of lactation. A high-quality feed and protein-energy supplement should be given to first-calf heifers because the cow herd is still expanding and giving birth to calves. For several months, calves graze on fodder and nurse on cows. They start to graze fodder from 4–5 weeks of age, and over the next few months, this becomes their main source of nutrition. A creep feeding arrangement is a management method that offers energy supplies other than milk together with, often, grass or hay. During the nursing stage, a calf that receives creep feeding typically gains more weight. Although the effects of creep feeding on calf performance can be observed practically year-round, the highest benefit may occur in situations where the quality and quantity of pasture or hay are subpar. Grain and protein supplements should form the foundation of creep feed. Replacement heifers and post-weaning calves are fed high-quality fodder on a free-choice basis. Grain and protein supplements can be added as needed to achieve the appropriate level of body weight gain. After they are weaned, calves may enter the feedlot right away or they may be raised on roughage for a year or longer before doing so. Roughages make up the majority of a stocker's diet, which is balanced with enough protein, minerals, and vitamins to support growth.

## 3.8. FEEDING SMALL RUMINANTS

Goats and sheep should be fed in accordance with their dietary requirements. Their dietary needs are influenced by a variety of characteristics, including breed, age, body weight, physiological stage, and yield level. The digestive efficiency of goats, cattle, and sheep is comparable. Goats are often seen to be superior browsers than sheep because they can better digest fiber, have a higher voluntary feed intake, and are fed lower- to medium-quality diets. Neonate lambs and kids with immature prestomach (rumen, reticulum, and omasum) have digestive systems that are remarkably similar to those of monogastric mammals. Except lactose or grain-based diets, they are unable to digest regular carbs. Colostrum is a newborn's first food, supplying all the nutrients and antibodies. Rumen gets colonized with microorganisms by the consumption of dry feeds, which include high-quality roughage and concentrates. The microorganisms encourage the pre-stomach's growth and development as they proliferate and start to break down food. The rumen of lambs and kids is typically active between 45 and 60 days of age. Lamb and kid may be grown on sheep and goat milk or milk substitute after receiving an acceptable amount of colostrum feeding. Producers may choose to utilize artificial raising for a variety of reasons, including low milk output, increased milk price, lower feed expenses, and so on. It is advised that liquid food feeding be continued in both rearing systems until the weight of the lamb and kid reaches at least 10 kg. They start eating solid foods at two weeks of age, and when pasture quantity or quality is scarce, they should be creep-fed. Ground or cracked maize, alfalfa hay or meal, soybean meal, oats, and molasses are common feed ingredients for creep rations. The creep ration ought to have no urea and between 18 and 20% crude protein. Feeding creep meals ad libitum or free choice can improve lamb and kid performance by promoting rumen growth. Sheep and goats have nutrient requirements that are slightly higher than maintenance during the early and middle stages of placental development. Due to the rapid growth of the fetus during the final 50 days of gestation, or the last trimester, they require significantly more nutrients, especially for ewes and goats carrying multiple fetuses. The ewes' and goats' peak milk production occurs three to four weeks after lambing and kidding. More milk is produced by ewes and goats that have twin or triplet lambs and kids than by those with singles. They should be fed premium forages supplemented with concentrates throughout the early lactation phase when their nutritional needs are at their highest. A concentrate ratio of between 50% and 60% is adequate. The amount of feed fed to each animal should be decreased after the first 60 days of lactation to prevent the accumulation of extra fat and to achieve the ideal body condition score.

Chapter 4

# Feeds Stuff Raw Materials

## Contents

## 4.1. OVERVIEW

To meet the nutritional demands of livestock, a wide variety of feed ingredients, often known as feedstuffs, are available. The raw materials used to create animal cells, tissues, organs, and products are these feedstuffs. But getting raw materials for animal feed from the farm to the feed mill and ultimately to the animal is a multi-step, intricate operation. Farmers grow crops like corn, soybeans, and wheat on the farm. After being harvested, these crops are moved to a feed mill to be turned into animal feed. The livestock is fed animal feed that is transported to the farm, giving them the nutrition they need to thrive and produce meat, milk, or eggs. Raw materials are subjected to a variety of circumstances during this travel, some of which may encourage microbial loads. The quality of the feed's constituent parts directly affects the feed's quality. However, consistency in ingredient quality is not always achieved, even with the finest procurement efforts. High microbial and pathogen loads in feed have been shown to lower animal performance and raise morbidity and mortality rates. Geographical location, seasonality, feed formulation, transportation, and storage conditions can all affect the microbial loads in feed and feed ingredients. In order to preserve the high caliber of feed materials, handling must be taken into account. Various Raw Materials for Animal Feed Types include:

- **Traditional Feed:** This word is typically used to refer to both household leftovers and the waste products from enterprises that process food. Pigs are fed food waste; chickens are fed scraps; wasted grain is a byproduct of the brewing industry; And a few common examples of conventional feeds are corn, soy, and peanuts.
- **Compound Feed:** These are compound mixes created for a certain animal's diet (blended feed). To optimize the benefits for the targeted animal, raw ingredients and extra additives are therefore designed to suit the requirements of the specific animal. Typically, the main ingredients of the same are fodder grains including corn, sorghum, soybean, oats, and barley.
- **Premixes:** These are occasionally sold separately or categorized as compound feeds. These are commonly available from companies that sell premixes in bags and mix micronized ingredients into commercial rations. The farmer can easily purchase the premix of his choice based on the animals he raises.

## 4.2. FEED COMPOSITION

Feedstuff composition varies. In contrast to chemicals that are "chemically pure," which means their composition is always the same, feeds' composition varies for a variety of reasons. Tabular data is not as accurate as an actual analysis of a feed to be utilized in a ration. Get real analysis and apply it whenever you can. Nevertheless, in many cases, true compositional data cannot be determined, or there is not enough time to obtain an analysis. The next best source of information is tabulated data. Keep in mind that feeds differ in composition while working with tabular data. Table values may differ by as much as 15% for the organic contents (such as crude protein, ether extract, crude fiber, acid detergent fiber, and neutral detergent fiber), up to 30% for the mineral elements, and at least 10% for the energy values.

1. **Dry Matter:** There is a wide range in the moisture content of meals. Therefore, the primary cause of fluctuation in feedstock composition on an "as fed basis" may be the DM content. Because of this, feed composition data is displayed on a dry matter basis, including both biological characteristics and chemical makeup. Ration formulation on a DM basis is sounder than using "as fed basis" because DM can vary substantially and since the DM content of feeds is one of the elements influencing total feed intake.
2. **Protein:** Calculating the Kjeldahl nitrogen multiplied by (100/16 or 6.25) yields the crude protein. (On average, proteins contain 16% nitrogen.) Crude protein does not disclose the true protein and nonprotein content of a meal. Although digestible protein has been included in many feed composition tables, it is more misleading than crude protein (CP) due to the significant contribution of body protein to the apparent protein in feces. Use the following formula to determine digestible protein from the CP content of the feed given to sheep or cattle: % digestible protein (DP) is equal to 0.9 (% CP)–3, where % CP and % DP represent the dry matter base ration values. The percentage of protein that enters the rumen and is not broken down by rumen microbes is known as "by-pass" protein or undegraded intake protein (UIP). These numbers are not constant, like other biological characteristics. For many feeds, the by-pass values are unknown, making it challenging to arrive at reasonable figures. Rumen microorganisms use degradable intake protein (DIP) to satisfy their nitrogen needs. The most affordable sources of DIP are those that contain nitrogen, such as urea. Meeting the ruminants' needs for metabolizable protein can be done more precisely by balancing the sources of DIP and UIP.
3. **Crude, Acid Detergent and Neutral Detergent Fibers:** Crude fiber (CF) is becoming less often used as a gauge of low-digestible material in feeds after more than a century of use. The main issue

with CF is that different feeds are subjected to varying degrees of indigestible lignin removal throughout the CF process. Even though it has been demonstrated that many feeds have a better digestibility for CF than NFE, the material extracted under the previous approach was referred to as nitrogen-free extract (NFE) and was believed to be more digestible than CF. Because CF necessitates the measurement of TDN, it was kept in the feedstuff analysis analytical scheme. An alternative analytical approach known as acid detergent fiber (ADF) and neutral detergent fiber (NDF) has been established by newer procedures. Animal digestibility is closely linked to ADF. The availability of net energy from digestible energy and the voluntary intake of the feed are related to NDF. Both metrics are more valuable than CF because they have a stronger correlation with expected animal performance. Furthermore, the CF content of feeds will be mostly ignored if TDN is substituted with other energy value measurements. CF will be removed as more comprehensive information about the ADF and NDF content of feeds is created.

4. **Minerals:** In the majority of feeding scenarios, the minerals calcium (Ca) and phosphorus (P) are significant. As the amount of concentrate in the diet rises or when complete protein is replaced with nonprotein nitrogen, potassium (K) becomes significant. Sulfur (S) also becomes important when the ration's nonprotein nitrogen content rises.

5. **Vitamins:** The only vitamin that is generally useful for feeding sheep and cattle is vitamin A. The amount of vitamin A and carotene in feeds is mostly determined by the maturity, harvesting conditions, storage duration, and storage conditions. As such, it is generally not a good idea to get all of your vitamin A from harvested meals. There will most likely be enough vitamin A if roughages with a nice green color are eaten, or if roughages are fed as immature fresh forages (like pasture).

6. **Energy:** The TDN method is now widely used to express the energy content of feeds for sheep and cattle. However, TDN has a number of technical issues. In some meals, there is a weak correlation between crude fiber and NFE digestibility. TDN further exaggerates the importance of roughages in animal production when contrasted with concentrates. Some have claimed that since energy cannot be expressed in pounds or percentages, TDN is an invalid energy metric. That being said, this is less of a critique of TDN's predictive value and more of a scientific argument. Digestible energy (DE) and

TDN have a consistent relationship, as demonstrated by numerous studies, with one pound of TDN corresponding to two Mcals of DE. Of course, multiplication can be used to determine DE. Twice the percentage of TDN content (2%). Technically, DE is preferable over TDN since it is quantified in calories. TDN use should eventually decline with increased focus on ADF and NDF as CF substitutes and direct measurement of DE using the bomb calorimeter. However, it should be clear that TDN and DE are exactly as good in predicting animal performance. With the creation of the California net energy system, interest in the application of net energy (NE) in assessing diets for sheep and cattle was rekindled. The primary explanation for this is the increased predictability of outcomes based on feed energy used for growth (NEg) or maintenance (NEm). Predicting feed intake and, consequently, the percentage of feed that will be used for growth and maintenance is the biggest challenge with applying these NE values. When creating rations, some people just consider the NEg values. The same argument that was leveled at TDN also applies to this, in that NEg overestimates the feeding value of concentrates in comparison to roughages. Some people choose to use the average of the two NE values, although this would only apply to sheep or cattle that consume twice as much as is necessary for upkeep. Using the NEm value plus a multiplier time the NEg value, all divided by one plus the multiplier, is the most precise method to use these NE values to create rations. The amount of feed intake above maintenance in relation to maintenance is known as the multiplier. For example, the NE value of the ration would be NE = [NEg + (10/8) (NEg)] / [i + (10/8)] if 700-pound cattle are projected to eat 18 pounds of feed, of which 8 pounds are needed for maintenance. The theoretical superiority of NE in animal performance prediction over DE or TDN is undeniable. That dominance disappears, though, if rations are made only using NEg. Therefore, NEm and NEg must be utilized in some combination if NE is to be used.

### 4.2.1. Alfalfa

Alfalfa is a perennial legume used as fodder that typically lives four to eight years, but depending on the variety and climate, it can live up to 20 years. Alfalfa is a common forage crop farmed all over the world for cattle. It is typically harvested as hay, but it can also be fed as green chop, processed into silage, or grazed. Of the common hay crops, lucerne often offers the highest feeding value. Less often, it is utilized as grazing. Alfalfa is frequently the highest-yielding forage plant when cultivated on well-adapted soils, but its main advantage is the combination of high yield per hectare and high nutritional quality. Because of its high protein content and easily digestible fiber, it is mostly used as feed for high-producing dairy cows. It is also used sporadically for beef cattle, horses, sheep, and goats.

For meat rabbits, lucerne hay is a common source of fiber and protein. Because dehydrated alfalfa and alfalfa leaf concentrates contain a high concentration of carotenoids, which effectively color egg yolks and body lipids, they are used in poultry diets to color eggs and meat. Alfalfa sprouts are also consumed by humans in sandwiches and salads. Commercially accessible forms of dehydrated lucerne leaf as a nutritional supplement include tablets, powders, and tea. Livestock grazing on fresh lucerne may experience bloating, thus caution must be exercised while managing this hazard. Its root nodules, like those of other legumes, are home to the nitrogen-fixing bacterium *Sinorhizobium meliloti*, which makes its meal high in protein regardless of the amount of nitrogen present in the soil. Its capacity to fix nitrogen, which raises soil nitrogen levels, and its application as animal feed significantly boost agricultural productivity. Alfalfa grows best on well-drained soils with a neutral pH of 6.8 to 7.5 and can be sown in the spring or fall.

For healthy growth, lucerne needs constant supplies of phosphate and potassium. It is still planted in the dry southwestern United States, where salinity is becoming a problem, but it is only somewhat sensitive to salt levels in the soil and irrigation water. Manure or chemical fertilizers can be used to improve low-fertility soils; however, pH correction is especially crucial. Generally speaking, seeding rates should range from 13 to 20 kg/ha (12 to 18 lb/acre), with variations depending on the region, type of soil, and technique used. A nurse crop can cause competition for light, water, and nutrients; however, it is occasionally used to minimize weed issues and soil erosion, especially for spring plantings. Alfalfa is harvested three to four times a year in most regions, but in Arizona and southern California, it can be harvested up to 12 times a year. In temperate regions, total yields are usually about 8 tons per hectare (3+1⁄2 short tons per acre), although yields as high as 20 tons per hectare (9 short tons per acre) have been reported. The location, the type of weather, and the crop's maturation stage at harvest all affect yields. Although the production is improved, later cuttings have less nutritional value (Table 4.1).

Table 4.1. *Typical Composition of Alfalfa*

| Items | ZC | ZM1 | ZM2 | GN | WL | XJD | SEM | *p*-value |
|---|---|---|---|---|---|---|---|---|
| DM (g/kg FM) | 463.66d | 464.33d | 473.65d | 497.22bc | 506.51ab | 513.53a | 5.10 | <0.001 |
| CP (g/kg DM) | 178.95bc | 166.62c | 167.76c | 176.12bc | 222.34a | 187.93b | 4.93 | <0.001 |
| WSC (g/kg DM) | 38.37c | 41.66c | 44.64c | 44.64c | 57.73b | 66.06a | 2.48 | <0.001 |
| NDF (g/kg DM) | 417.51ab | 411.17abc | 419.94a | 382.35abcd | 364.79d | 387.98abcd | 6.98 | 0.1 |
| ADF (g/kg DM) | 345.87a | 308.08ab | 287.59bc | 257.24c | 266.52c | 267.7c | 8.60 | 0.003 |

Values with different letters indicate significant differences between fresh materials. SEM, standard error of the mean; DM, dry matter; CP, crude protein; NDF, neutral detergent fiber; ADF, acid detergent fiber; WSC, water-soluble carbohydrate; ZC, Zhongcao No. 3; ZM1, Zhongmu No. 1; ZM2, Zhongmu No. 2; GN, Gongnong No. 1; WL, WL168; XJD, xinjiangdaye.

| **Feed-stuff** | **DM (%)** | **CP (%)** | **By-Pass (%)** | **EE (%)** | **CF (%)** | **ADF (%)** | **NDF (%)** | **Ash (%)** | **Ca (%)** | **P (%)** | **K (%)** | **S (%)** | **ZnP PM** | **TDN (%)** | **DE Mcal/ lb.** | **NEm Mcal/ lb.** | **NEg Mcal/ lb.** |
|---|---|---|---|---|---|---|---|---|---|---|---|---|---|---|---|---|---|
| Alfalfa cubes | 91 | 18 | 35 | 2.0 | 29 | 34 | 45 | 11 | 1.3 | 0.23 | 1.9 | 0.35 | 18 | 57 | 1.14 | 0.56 | 0.25 |
| Alfalfa dehy-drated 17% | 92 | 19 | 60 | 3.0 | 26 | 35 | 45 | 11 | 1.4 | 0.25 | 2.7 | 0.24 | 19 | 61 | 1.22 | 0.61 | 0.31 |
| Alfalfa fresh | 26 | 19 | 20 | 2.2 | 27 | 32 | 44 | 8 | 1.6 | 0.32 | 2.3 | 0.34 | 21 | 60 | 1.20 | 0.59 | 0.30 |
| Alfalfa hay early bloom | 90 | 18 | 20 | 2.2 | 29 | 35 | 47 | 8 | 1.4 | 0.25 | 2.3 | 0.30 | 18 | 60 | 1.20 | 0.59 | 0.30 |
| Alfalfa hay mid-bloom | 89 | 17 | 25 | 2.0 | 30 | 38 | 50 | 10 | 1.4 | 0.23 | 1.8 | 0.30 | 17 | 58 | 1.16 | 0.57 | 0.26 |
| Alfalfa hay full bloom | 88 | 16 | 30 | 1.8 | 34 | 41 | 56 | 8 | 1.3 | 0.20 | 1.7 | 0.29 | 17 | 53 | 1.06 | 0.52 | 0.18 |
| Alfalfa hay mature | 90 | 14 | 35 | 1.7 | 38 | 45 | 59 | 8 | 1.3 | 0.19 | 1.4 | 0.25 | 17 | 50 | 1.00 | 0.49 | 0.12 |
| Alfalfa silage | 30 | 18 | 20 | 3.0 | 30 | 35 | 46 | 8 | 1.5 | 0.28 | 2.4 | 0.30 | 17 | 54 | 1.08 | 0.53 | 0.20 |
| Alfalfa silage wilted | 36 | 17 | 25 | 3.0 | 30 | 35 | 46 | 8 | 1.5 | 0.28 | 2.4 | 0.30 | 17 | 58 | 1.16 | 0.57 | 0.26 |

*Source: Lin, J., Li, G., Sun, L., Wang, S., Meng, X., Sun, L., Yuan, L., & Xu, L. (2023). Varieties and ensiling: Impact on chemical composition, fermentation quality and bacterial community of alfalfa. Frontiers in Microbiology, 13, 1091491. https://doi.org/10.3389/fmicb.2022.1091491.*

Alfalfa is typically trimmed and baled before being used as hay. While bales are easier to utilize for transportation, storage, and feed, loose haystacks are still in use in some places. Since carbohydrates are at their peak at the one-tenth bloom, the initial cutting should ideally be taken at the bud stage, and the successive cuttings should be taken just as the field is starting to flower. A swather splits and organizes lucerne in windrows when harvesting by hand; when utilizing farm equipment, this process is done manually. Hay is cut with a device called a mower-conditioner in regions where the lucerne does not quickly dry out on its own. The lucerne dries more quickly with the mower-conditioner because of its set of rollers or flails, which crimp and snap the stems as they go through the mower. Hay is collected into bales by a tractor towing a baler once the lucerne has dried. For lucerne, a variety of bale types are frequently utilized. Alfalfa is baled into small, two-string bales for small animals and individual horses. These are named after the string strands that are used to wrap the alfalfa. Up to half-ton (six-string) "square" bales, which are actually rectangular and measure approximately 40 cm × 45 cm × 100 cm (16 in × 18 in × 39 in), there are other types of bales available, including three-string. Depending on the moisture content, little square bales weigh between 25 and 30 kg (55 and 66 lb) and are readily broken into "flakes" by hand. Large round bales, usually measuring 1.4 to 1.8 m (4+1⁄2 to 6 feet) in diameter and 500 to 1,000 kg (1,100 to 2,200 lb) in weight, are used in cattle ranches. These bales can be unrolled on the ground for huge herds of cattle or stacked in stables or large feeders for horse herds. The bales can be handled with a grapple (claw) on the front-end loader of the tractor, or they can be loaded and stacked using a spike, called a bale spear, which pierces the center of the bale. Alfalfa is frequently turned into haylage for dairy cattle by a procedure called ensiling. The lucerne is finely chopped and fermented in silos, trenches or bags where the oxygen supply can be restricted to facilitate fermentation, as opposed to being dried to make dry hay. Alfalfa's anaerobic fermentation process keeps its high nutritional levels comparable to those of fresh fodder, and it tastes better to dairy cows than dried hay. Alfalfa silage is frequently infected with several microbial strains to enhance the silage's aerobic stability and fermentation quality.

### 4.2.2. Ammonium Sulfate

Ammonium sulfate ($(NH_4)_2SO_4$) is an inorganic salt with a variety of industrial applications, most commonly used as a soil fertilizer. It contains 21% nitrogen and 24% sulfur. Sulfur is a crucial component of proteins, certain vitamins, and several important hormones, playing a key role in animal nutrition. Sulfur is essential for various bodily processes, including protein synthesis and metabolism, lipid and carbohydrate metabolism, blood coagulation, endocrine function, and the maintenance of acid-base balance in intra- and intercellular fluids.

In ruminant diets, where natural protein levels may be suboptimal, non-protein nitrogen sources—typically urea—are used to balance ammonia and total protein levels. However, animals on high-grain diets containing urea may have adverse reactions to excessive sulfur. Therefore, it is essential to adjust all ruminant diets for specific nutrients, particularly ensuring the proper protein-to-sulfur ratio in supplemental feeding regimens.

Specifications for ammonium sulfate:

- **Nitrogen (N), minimum**: 21.00%;
- **Equivalent crude protein from non-protein nitrogen, minimum**: 131.25%;
- **Sulfur (S), minimum**: 24.00%;
- **Screen Analysis – Tyler**: +10 / 0.5%, +28 / 100%;
- **Bulk Density**: 67–71 lbs/ft$^3$;
- **Color**: White;
- **Moisture**: 0.2%.

### 4.2.3. Barley

Hordeum Vulgare, an annual grass, is the source of barley (*Hordeum* spp.), a cereal. This versatile grain should be the first choice on the farm for feeding cattle. Typically, barley grain is provided in addition to forage. Barley can, however, be fed to appetite in situations when the goal is extremely quick daily liveweight increase, such as in cereal beef systems. Introduce 0.5 kg/head increments every other day for a minimum of 10 days to ensure a gentle introduction. If it is being fed ad lib, fresh feed should always be added to the troughs or hoppers, and the ration should be balanced in terms of protein, minerals, and vitamins. Depending on the animal's liveweight, a source of fiber in the form of long, clean cereal straw should be given when barley is supplied to appetite. The recommended intake should be no more than 13% of the daily millet (up to 2 kg/head). Don't rely on livestock eating enough bedding; provide feeders with straw instead. Selenium and vitamin E supplements are crucial when feeding moist cereals since the fermentation and oxidation process reduces these nutrients in wet grain. For feed barley, no particular quality requirements exist. Barley that doesn't match food or malt criteria is frequently turned into animal feed; this is known as covered barley. It could be necessary to mechanically shatter the pericarp and hull of covered barley kernels in order to boost nutritional absorption in monogastric animals and regulate fermentation rates in large ruminants. Pelletizing, crushing, flaking, steaming, and cracking are a few examples of processes. Certain types of barley, such hooded barleys, are grown expressly for animal feed (Table 4.2).

**Table 4.2.** *Chemical Composition of Covered and Naked Barley Grain*

| Chemical Composition (%) | Covered Barley | Naked Barley |
|---|---|---|
| Starch | 57.7 | 60.7 |
| Protein | 12.2 | 15.1 |
| Fat | 2.5 | 2.7 |
| Sugars | 1.2 | 1.5 |
| Ash | 2.1 | 1.6 |
| Dietary fiber | 20.6 | 16.6 |
| β-glucan | 4.8 | 5.7 |

*Source: Lukinac, J., & Jukić, M. (2022). Barley in the production of cereal-based products. Plants, 11(24), 3519. https://doi.org/10.3390/plants11243519.*

Perfect for use as ruminant feed, with an average protein content of 11–12% and a range of 6–14%. Its high energy content, which comes from starch, makes it highly fermentable, promoting milk protein and rapid growth, which is advantageous to ruminants. About 10% of barley can pass through the rumen unfermented, but like all cereals, the protein is of average quality and is especially lacking in lysine. While barley is a great addition to forages, it requires careful mineral and vitamin balance, especially when it comes to treated grains. It is particularly deficient in calcium, vitamins D, E, and A. The vitamin E content is significantly decreased when damp grains are processed with propionic acid.

## 4.2.4. Beans

Even though producers have access to a wide range of feed ingredients, protein remains a highly sought-after nutrient. This isn't necessarily because it's hard to get, but rather because the price of protein is constantly rising. The plant Vicia faba, also known as the "faba bean," is one possible remedy. In addition to these common names, this bean is also known by the names "horse bean," "field bean," and "broad bean," which is the Italian translation of the word "fava" and likely the most similar sounding term to "faba bean." Because faba beans don't contain much oil, livestock producers typically plant them as a domestic source of protein that can be digested quickly on the farm. They also don't require roasting because they don't contain anti-nutritional enzymes. Related to lima beans, faba beans have a protein content of about 30%. They offer a reliable source of starch-based energy. The majority of the cultivars that are sold in Canada are grown in the Netherlands and Germany. It was previously recognized that early varieties had an unfavorable amount of tannin, primarily in the hull, which decreased the digestion of protein. Monogastric animals cannot absorb the tannins, even though they give the bean more taste and are digested by ruminants. Varieties have evolved over time to drastically lower tannin content. If the plant produces a purple or pink blossom during the flowering stage, it is easy to determine whether the variety contains tannins. Usually, the food sector uses these. Imposa, Snowbird, and Tabasco are among the kinds with a white flower that have very little or no tannin content (Table 4.3).

**Table 4.3.** *Compositional Analysis of Faba Bean Fractions, Average Contents ± Standard Deviation (g/100 g DM)*

| | PI 1 | PI 2 | SF | FRF |
|---|---|---|---|---|
| **Protein** | 72.64 ± 0.31 [a] | 71.37 ± 0.93 [a] | 0.18 ± 0.01 [c] | 16.70 ± 0.08 [b] |
| **Fat** [1] | 6.24 ± 0.18 [a] | 0.78 ± 0.04 [b] | n.d. | n.d. |
| **Digestible Starch** | 5.32 ± 0.28 [c] | 1.00 ± 0.07 [d] | 28.46 ± 0.37 [a] | 15.06 ± 0.22 [b] |
| **Resistant Starch** [2] | 0.33 ± 0.01 [c] | n.d. | 55.41 ± 3.31 [a] | 9.70 ± 0.44 [b] |
| **Σ Total Starch** | 5.65 ± 0.27 [c] | 1.00 ± 0.07 [d] | 83.87 ± 3.07 [a] | 24.76 ± 0.61 [b] |
| **Total Dietary Fiber** | 8.69 ± 0.67 [c] | 7.64 ± 1.53 [c] | 62.28 ± 0.76 [b] | 69.09 ± 0.24 [a] |
| **FODMAPs** [3] | n.d. | 0.07 ± 0.00 [a] | 0.06 ± 0.00 [a] | n.d. |

*Sugars were not detected or were at levels below 0.02 g/100 g DM. Different letters indicate significant difference ($p < 0.05$).*

[1] *n.d.: not detected.*

[2] *n.d.: not detected or below 0.09 g/100 g DM.*

[3] *n.d.: not detected or below 0.025 g/100 g DM.*

*Source: Krause, M., Sørensen, J. C., Petersen, I. L., Duque-Estrada, P., Cappello, C., Tlais, A. Z. A., Di Cagno, R., Ispiryan, L., Sahin, A. W., Arendt, E. K., et al. (2023). Associating compositional, nutritional, and techno-functional characteristics of faba bean (Vicia faba L.) protein isolates and their production side-streams with potential food applications. Foods, 12(5), 919. https://doi.org/10.3390/foods12050919.*

To achieve maximal nitrogen fixation, it is recommended to inoculate faba beans in order to promote a more uniform distribution and faster nodule development. Their exceptional capacity to fix nitrogen makes them a useful crop, much like other pulses, since it reduces the amount of nitrogen required for the next crop. To encourage the growth of their seedlings and the development of their root system, faba bean plants require a significant amount of phosphorus. If necessary, phosphate can be banded with the seed based on the amount of phosphorous that is readily available in the soil. Planting should generally be completed by the end of the first week of May; some farmers put the deadline for planting on May 7. Because they can withstand frost, beans should be planted when the soil temperature is approximately 3°C. It has been demonstrated that delayed planting lowers yield because of drought stress and late maturation when high temperatures are encountered during flowering. Variations in planting timing of up to two weeks can result in yield reductions of up to 32% (Table 4.4).

**Table 4.4.** *Laboratory Analysis of Faba Beans: Techno-Functional Properties of Faba Bean Fractions, Values ± Standard Deviation*

| | | **PI 1** | **PI 2** | **SF** | **FRF** |
|---|---|---|---|---|---|
| WI | (%) | 81.21 ± 0.95 [c] | 88.10 ± 0.21 [b] | 96.52 ± 0.50 [a] | 80.21 ± 0.74 [d] |
| pH | (–) | 5.14 ± 0.04 [c] | 4.57 ± 0.00 [d] | 8.02 ± 0.06 [a] | 6.36 ± 0.02 [b] |
| TTA | (mL/g) | 4.80 ± 0.05 [b] | 6.67 ± 0.55 [a] | 0.02 ± 0.00 [d] | 0.45 ± 0.01 [c] |
| FC | (%) | 18.06 ± 2.41 [b] | 133.33 ± 15.73 [a] | 0.00 ± 0.00 [d] | 7.92 ± 1.34 [c] |
| FS | (%) | 70.00 ± 8.66 [a] | 39.79 ± 4.09 [b] | – [2] | 0.00 ± 0.00 [c] |
| WBC | (%) | 149.80 ± 0.61 [b] | – [1] | 63.67 ± 1.28 [c] | 1028.92 ± 56.90 [a] |
| OBC | (%) | 65.32 ± 0.33 [c] | 229.13 ± 4.42 [b] | 49.43 ± 1.55 [d] | 342.35 ± 4.96 [a] |

*Abbreviations: WI: Whiteness index; TTA: total titratable acidity; FC: foaming capacity; FS: foam stability; WBC: water binding capacity; OBC: oil binding capacity; results referred to DM. Different letters indicate significant differences ($p < 0.05$).*

[1] *No analysis of water binding capacity was carried out on PI 2 due to high solubility.*

[2] *No analysis of foam stability due to no foaming capacity.*

*Source: Krause, M., Sørensen, J. C., Petersen, I. L., Duque-Estrada, P., Cappello, C., Tlais, A. Z. A., Di Cagno, R., Ispiryan, L., Sahin, A. W., Arendt, E. K., et al. (2023). Associating compositional, nutritional, and techno-functional characteristics of faba bean (Vicia faba L.) protein isolates and their production side-streams with potential food applications. Foods, 12(5), 919. https://doi.org/10.3390/foods12050919.*

### 4.2.5. Beet

The biennial fodder beetroot (*Beta vulgaris* subsp. *vulgaris* L.) is produced for its plump, fleshy roots. Although it can also be fed to pigs, ruminants are the primary consumers of fodder beetroot. Fodder beetroot is a vital source of energy and is appealing due to its high sugar content. Grown specifically to replace grains as a more cost-effective option for dairy and beef cattle feed, feed beets are sugar beets. The cost per acre to grow feed beets is comparable to that of maize, but the dry matter yields are significantly higher than corn grain. Beyond its economic value, sugar beets are a sustainable crop that can withstand drought, improve soil quality, and be cultivated in a variety of soil types.

Feed beets are also an adaptable feed component. They can be chopped and added to other ensiled feeds like haylage or corn silage. They can be fed fresh (as they store well in piles) or ensiled. However, due to their high moisture content, they are most effective when used locally. Understanding the method used to extract sugar from sugar beets is crucial to understanding the nutritional properties of sugar beet byproducts.

Before processing, extraneous materials such as small beetles and leaves are removed at the processing facility. This material, known as "sugar beet tailings," is also fed to animals. The beets are sliced into long strips called cossettes, which are boiled in hot water to remove the sugar—a process known as diffusion. The sugar and hot water mixture is further processed to produce bulk or bagged sugar after diffusion is complete.

The cossettes are then pressed to remove part of the pulp's water content, making transportation easier. Pulp intended for sale as dry shreds or pellets is dried to around 10% moisture content in a pulp dryer. Pelletizing the pulp often follows drying to facilitate transportation and storage. The refining process also produces molasses, with the amount typically ranging from 4% to 5% of the raw sugar beet weight. Centrifugation is used to separate the molasses from the sugar-containing juice. Molasses contains 40% to 50% residual sugar.

Molecular exclusion chromatography is commonly used to further refine molasses, producing desugared molasses, also known as condensed separator byproduct (CSB). This process removes some of the remaining sugar and concentrates the remaining nutrients.

- **Sugarbeets:** Processors often have to throw away entire sugarbeets because of rotting or overly big harvests. Cattle can be effectively fed whole beets. Whole beets have a low crude protein content of 6.8% but a high total digestible nutrition content of 75% to 81%. Before feeding, entire beets should ideally be split up. Growers can break up whole beets by using longer mixing durations with a traditional mixer wagon. Furthermore, some farmers indicate that whole beets may be successfully placed on stubble or stalk fields using manure spreaders and that cows can access the beets on the field. When giving entire beets to young ones, choking could be an issue. To lower the risk of choking, whole sugarbeets can also be fed chopped in a forage harvester or tub grinder. Before feeding, some producers have found success by chopping the beets with a forage harvester.

- **Beet Pulp:** Beetroot pulp is a useful supplement for cows that are pregnant or nursing, a component in diets for backgrounding, or a substitute for roughage in diets for finishing. Crude protein content in beetroot pulp is comparatively low (8%), whereas TDN content is rather high (72%). Wet beetroot pulp, with an energy value higher than corn silage, can be employed as the roughage source in finishing diets. It can be added to finishing rations at a rate of 5% to 15% of the diet (DM basis) as a partial or full roughage replacement. Cattle will eat 30 to 35 pounds of wet pulp when wet beetroot pulp is offered at 20% of the diet dry matter (DM), which is less than 1% of their body weight.
- **Beet Tops:** Since beetroot tops are practically mulched when they are removed, their utility has been limited by advancements in defoliator technology. This makes it difficult for animals to graze the material or to collect the residual residue into a windrow.
- **Beet Tailings:** Beet tailings are made up of dirt, broken or damaged beets, small beets, and other foreign materials that are not good for making sugar. Tailings have a high moisture level (75%–85%) and a very variable nutritional content. Beet tailings, with minimal amounts of foreign material and soil contamination, provide a feeding value that is comparable to or somewhat higher than maize silage on a dry-matter basis. On the other hand, a large amount of dirt or foreign matter will significantly lower the energy content of tailings. Beet tailings may cause choking hazards. With beetroot tailings, shipping is a significant cost because of the high moisture content. Before acquiring these leftovers, producers should carefully consider the expenses associated with transportation. Beet tailings have been successfully ensiled by some producers as a storage solution. A dry feedstock should be combined with tailings to reach a moisture content that prevents the silage pile from losing too much effluent. The addition of chopped forages, grain screens, or other byproduct feeds can help achieve this.

- **Molasses:** The main use of beet molasses (75% TDN) in animal feed is as an energy source; however, it is also added to rations to enhance palatability and minimize dust. It is also added to manufactured feeds as a pellet binder and a transporter for vitamins, minerals, and urea or other nonprotein nitrogen sources. The highest percentage that is advised is 15%. When compared to cereal grains, beet molasses has comparatively higher quantities of potassium, sulfur, magnesium, sodium, and chloride. Beet molasses needs to be kept in a warm, well-insulated tank. It can be readily pumped at temperatures exceeding 110°C with a 2.5–3 horsepower pump and 3-inch tubing.

### 4.2.6. Corn

Both domestically and internationally, maize products play a significant role in our daily lives. However, did you also know that it forms the basis of many livestock animals' diets? Field corn is fed to animals by farmers for a number of reasons, including its low cost and high nutritional content, which is essential for healthy growth and development. It is understandable why maize is so widely used as animal feed given all of its advantages! Corn grain is the primary raw material used to make animal feed; it makes up between 50 and 70% of the diets of animals that are monogastric (mostly pigs and poultry).

### 4.2.7. More Feedstuff Components

- Diammonium phosphate;
- Dicalcium phosphate;
- Distillers' products;
- Garbage;
- Grass;
- Hop;
- Meat meal;
- Oat;
- Fruits;
- Pulses;
- Tubers;
- Manure;
- Sorghum;
- Wheat.

## 4.3. EUROPE CO-PRODUCTS IN ANIMAL NUTRITION

Farm animals that generate food, particularly ruminants like cattle and sheep, have the special ability to retain the byproducts of processing arable items in the food chain. These byproducts are either not used by humans as food or drink, or they are used to make biofuels or other industrial products. Because of this, the livestock industry plays a crucial role in the circular economy. Many people are unaware of these aspects of agricultural production, and this benefit is typically disregarded. Usually, the process that produces the final feed materials is one in which the primary output is a separate consumer good, including food, drink, biofuels, or other industrial uses. Despite being an inevitable byproduct of the lead process, they are referred to as co-products because of their great value. Using co-products to their full potential helps the food production system remain profitable and sustainable. Co-products are related to the production of all food types, including dairy, meat, and vegetable-based goods. Utilizing animal nutrition technology, these byproducts can be made into high-performing, nutritious feedstuffs for animals that produce food by extracting their nutritional and financial worth. The procedure entails deftly examining the nutritional elements included in co-products and comparing them to the dietary and physiological needs of particular farm animals at particular phases of their lives. It involves regulating any naturally occurring compounds, known as "anti-nutrients," that might be detrimental to an animal's health or performance while also making sure that the taste, safety, and quality of animal products produced and purchased by consumers are not compromised and, in some cases, even improved. The utilization of co-products aligns with the core principles of the compound feed sector, which pinpoint the nutrients present in a wide range of safe individual feed ingredients, including cereal grains, pulses, and co-products, and provides the most economical feed resources by repurposing components that would have otherwise been wasted. The use of co-products in the feed market not only increases the economic worth of the original feed material but also helps to maintain the environmental sustainability of the food chain. Simultaneously, it increases the livestock agricultural industry's competitiveness by making alternative, reasonably priced feed supplies more widely available. Naturally, the nutritional integrity and safety of co-products must be maintained by producers during the production, handling, storage, and transportation processes in order for them to have any use as feedstuffs. Stated differently, considering them as "products" rather than as throwaway "waste." The goal of FEFAC is to support and motivate co-product providers to manage their manufacturing operations in a way that best preserves and even improves the nutritional content of the materials intended for feed. Few other industries detect and reuse resources as a byproduct of other manufacturing processes, most of which are directed towards other high-end products, with the rigor that the compound feed industry does. The industry prides itself on being a "resource efficiency champion." For instance, brewers' grains are a byproduct of the manufacture of beer. Although they are useless for human consumption, they are now a lucrative feed ingredient because of advances in animal nutrition. Hopefully, when you use the consumer products you see in this publication, you will remember that the European compound feed industry was there to make the best use of the co-products. The industry is working hard to raise awareness about the role it plays in the circular economy.

### 4.3.1. Wheat and Rice Milling

It is fitting to start with the byproducts of milling wheat, since here is where the European compound feed industry had its start in the early 1900s. Indeed, the word "feed mills" originates from the fact that compound feed manufacture was first done by flour millers themselves. Though once regarded as a waste product, surplus particles from grinding grains to make flour are now saved and used as co-products for feed, helping one of the largest industrial food ingredient sectors in its manufacturing process. Only the endosperm (flour) is used in the making of non-whole-wheat bread, breakfast cereals, and pasta; the hard outer layer (bran) is left behind. The definition and composition of wheat bran might vary based on the place in which it is produced. Typically, flour millers maintain strong connections with adjacent feed markets, so bolstering local or regional supply chains. Since wheat bran is used to make whole-meal bread, most people are aware that it contains more protein (14–19%), fiber, minerals (especially calcium and phosphorus), and oils than endosperm in bread manufacture. The feed sector benefits from these nutritional characteristics as well. All kinds of farm animals can eat wheat bran; however, because of its high fiber content, poultry should only be fed a limited amount of it. Wheat bran is a very popular ingredient in feed for ruminants and sows. In the latter case, it is linked to increased milk fat output. Bran surrounds rice grains as well, however it is far less common for humans to eat than wheat bran. This is in spite of assertions that, while making up only 10% of the grain, the bran contains about 80% of the nutrients. Brown rice is turned into white rice by removing the rice bran from the endosperm. It has a respectable quantity of protein and fiber, just like wheat bran. Defatted rice bran is frequently used as a feed ingredient since the oil in it typically still plays a part in food production. Additionally, B and E vitamins as well as other trace minerals like zinc and manganese can be found in rice bran. Because of its high fiber content, rice bran is especially good for dairy animals, just as wheat bran.

## 4.3.2. Sugar Production

Sugar cane and sugar beetroot are the sources of sugar that are used as table sugar and as an ingredient in food and beverage items. The world's largest producer of beetroot sugar is the European Union. This is refined by slicing it into thin strips or pieces, and water heated to around 70°C is used to extract the sugar as juice. The residual beetroot pulp provides exceptional feed value for various livestock species and is high in calories and fiber. It is most frequently utilized in dairy feed because of its capacity to raise fat levels and milk yield while lowering the risk of rumen acidosis, which is brought on by an excess of grain starch. Beetroot pulp, after drying, is shipped or transported by trucks on a worldwide scale. Both molasses- and non-molasses-infused dried beetroot pulp pellets are available for purchase.

Molasses is a syrupy byproduct that is created when the sugar juice is further refined and crystallized. Although it is high in some minerals and offers a quick energy boost, molasses has other uses outside of food. It is a highly prized, energy-dense flavor enhancer that improves the uniformity and palatability of compound feed.

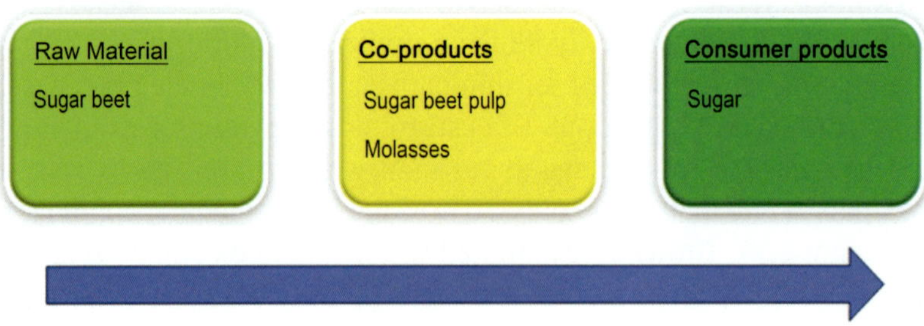


Molasses's viscosity renders it unfit for direct animal feeding; nonetheless, its excellent pellet-binding properties help regulate the numerous tiny particles involved in compound feed processing, averting post-production pellet degradation. In addition to being a major ingredient in ruminant feed, molasses is also occasionally added to pig and, to a lesser degree, poultry feed. Both beetroot and cane molasses are nutritionally equivalent; the key distinction is that beetroot molasses is made in Europe, whereas cane molasses is usually imported from outside. Beet pulp that has been pressed is also offered in fresh form. This can be ensiled (conserved by placing in a silo or silage clamp) or fed directly to animals. It is an excellent feed for all types of ruminants.

### 4.3.3. Beer Brewing

Beer's primary components are yeast, hops, malted barley, and water. The soluble portion of the malted barley is dissolved in water during the brewing process to produce wort, which is then fermented to produce beer. When farms and monasteries in Europe brewed their own beer before the Industrial Revolution, they gave the by-products to their cattle.

This leftover solid residue is made up of spent brewers' grains. Brewery by-products account for 20–24% of the original input in the almost 40 billion liters of beer brewed in Europe annually, therefore their use as feed is substantial. Brewers' grains are a great source of protein and highly digestible fiber for cattle and other ruminants, making them an ideal feed ingredient. The fiber enhances the performance of forage-based diets that are heavy in starch and low in easily fermentable fiber by supporting the rumen, the largest chamber of a ruminant's stomach. Brewers' grains are another excellent source of protein, with a crude protein level that ranges from 19 to 31% on a dry matter basis. Livestock near a brewery is also preferred because of their high moisture content, which makes them perishable.

The ingredient in beer brewing that turns carbohydrates and sugars into alcohol is called brewer's yeast. Once the yeast is removed after the brewing process, it can be used as animal feed after being deactivated by heat treatment or organic acids. Brewers' yeast has a beneficial amino acid profile that is similar to soya and is rich in protein (36–50% on a dry matter basis). It is a very adaptable feed that may be used with any kind of farm animal. It is acknowledged as an excellent source of B vitamins when added to chicken feed. A brewer's primary goal is to produce high-quality beer, but waste reduction and environmental effect mitigation are equally important. The brewing industry's continuous improvement of its environmental performance depends on ensuring that secondary products, such as brewers' grains and brewers' yeast, have a sustainable use as high-protein, high-fiber, and high-vitamin animal feed.

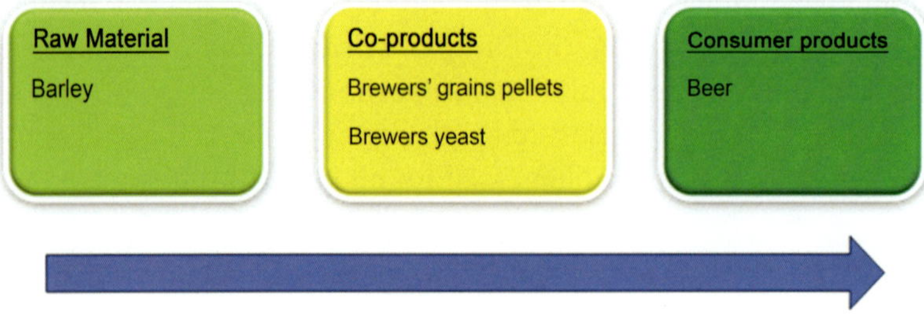


## 4.3.4. Starch and Ethanol Production

High-value products like starch are used in many industries, such as the food, pharmaceutical, fuel, paper, and textile industries. Both wet and dry milling techniques can be used to create it. Cereals such maize, wheat, barley, and rye, as well as starchy potatoes cultivated specifically for this use, are milled to separate the starch, fibers, and proteins. At different stages of the milling process, several feed co-products are generated. Depending on the kind of cereal, wet milling separates and purifies the grain's major constituents, which include protein, starch, fiber, soluble, and oil. Three significant feed co-products are produced by wet milling maize: maize gluten feed, maize gluten meal, and maize germ meal. Medium-protein, medium-energy feed is maize gluten feed, which is typically combined with maize germ meal. Due to its comparatively high digestible fiber content, it is frequently used in ruminant diets. Crude protein makes up 60–75% of maize gluten meal. Methionine is a high form of an important amino acid, while lysine is found in lower amounts than in soya. Wet and dry milling of wheat yields wheat gluten feed, wheat gluten meal, wheat germ meal, and wheat gluten itself. Bran, gluten, and maybe wheat germ are the ingredients of wheat gluten feed. It is a nutrient-dense protein and starch-rich, high-fiber component used in ruminant, pig, and poultry feed. With 75–80% crude protein, wheat gluten meal is a concentrated form of wheat protein that has a high digestion. The primary ingredient in wheat germ meal is wheat germ, along with a small amount of bran and "middlings" or "shorts." It has 9% fat and at least 25% protein. It has important fatty acids, vegetable oil, and a high digestible protein content. A great source of vitamins B1 and E is wheat germ meal. Even more protein—80% more—can be found in wheat gluten, which is utilized in aquafeed and in calf milk substitutes in hydrolyzed form. Potato protein and pulp are produced during the starch-processing of potatoes. Rich in the vital amino acids cystine, methionine, and lysine, potato protein is a premium and easily digested source of protein that is very beneficial for piglet diets. Dried potato pulp offers monogastric and ruminant animals an easily digested source of fiber. Although it is most commonly employed as a component of spirits like vodka, ethanol is also being utilized more and more as a biofuel. The "spent grains" that remain after the fermentation process used to extract ethanol from cereals are known as distiller's grains. Dried distillers' grains (DDGs) are a highly prized feed co-product that are abundant in soluble sugars, fat, protein, and fiber. They are traded all over the world. While maize is the primary source of DDGs, they can also be made from wheat, barley, rye, or a combination of all three, occasionally, contingent upon the price of cereal. Although the production method and the grain used determine the nutritional qualities and characteristics of DDGs, they are generally appropriate for all kinds of farm animals. In addition to being a rich supply of potassium, zinc, and phosphorus, DDGs' fiber content helps lower ruminal acidity in high-grain diets.

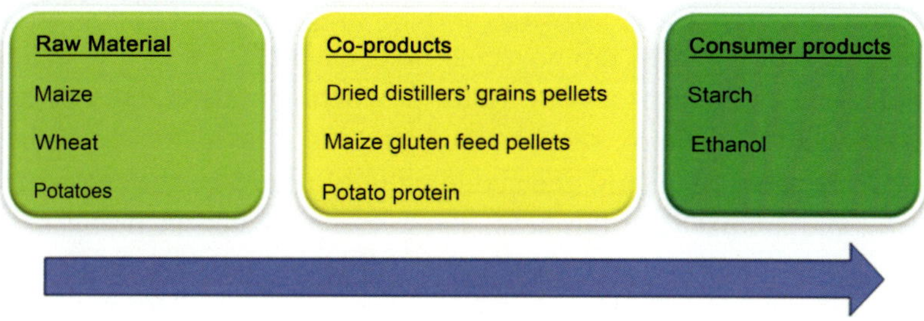


## 4.3.5. Citrus Fruit Processing

Worldwide, billions of liters of citrus fruit juice, including grape, orange, and lemon, are drunk. There is a solid residue consisting of the peel and seeds after the juice has been extracted. Known as citrus pulp, this is an important co-product for feed. Although fresh pulp can be given to animals on a local basis, it is normally pressed and dried into pellets in the fruit pressing facility itself before being shipped all over the world as a commodity (mostly from Brazil and the US). Citrus pulp has a high energy and fiber content and is easily digested by ruminant animals. Similar to sugar beetroot pulp, its highly digestible fiber content encourages good rumination in the stomachs of cows, resulting in copious amounts of saliva that act as a buffer on the rumen's pH. Orange-based citrus pulp adds a pleasant, fragrant flavor to the feed, which improves palatability. A higher concentration of lemon or grapefruit pulp in the pulp usually results in a more bitter flavor. Citrus pulp is known to preserve milk quality in dairy cattle, particularly milk fat. It is especially well adapted to ruminants, who can process high-fiber diets and serves as an excellent illustration of how an extra component of agricultural crop output can be retained in the food chain. Citrus pulp is infrequently utilized in poultry because of its high fiber content and the anti-nutrient limonin it contains, making it far less suited for pigs.

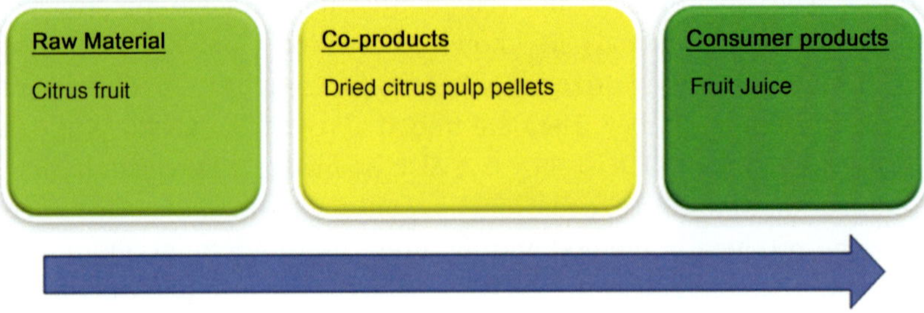


## 4.3.6. Industrial Fermentation Biomass

Microorganisms including yeasts, fungi, and bacteria are used in industrial fermentation to produce a growing variety of goods. Vitamins, enzymes, organic acids, carotenoids, amino acids, and alcohols are a few examples. Industrial fermentation is necessary for the production of detergents, cosmetics, medications, and bioethanol; some of the byproducts are also used in the food industry. The microorganisms' metabolic activity produces the ingredients of interest for the animal feed business. These are produced in liquid or solid-state bioreactors under tightly regulated conditions, and they are fed with a supply of nitrogen and carbohydrates like sugar or molasses. The main products are either generated inside the microorganisms and taken out of their cells, or they are expelled by them into the media in which they are isolated and concentrated. The microbial biomass and spent medium that are still present after the separation procedure are both highly valuable as feed. In comparison to chemically based processes, these biological processes offer the needed end products and associated coproducts under extremely sustainable and circular circumstances, while also saving a substantial amount of energy and resources. A burgeoning industry that is centered on manufacturing microbial biomass from diverse microorganisms and sustainable carbon sources, rather than targeted metabolites, is being spurred by the effectiveness of industrial fermentation. Single cell proteins (SCP) are the resultant product of this procedure as well as the co-products generated. They are made out of protein, either refined or crude, which is obtained from the cells of microorganisms that are cultivated on different carbon sources for synthesis, such as bacteria, fungus, yeast, and algae. SCP is high in lipids and vitamins and has a crude protein content of more than 40%. One of the greatest sources of proteins and amino acids is fermentation biomass; it contains over 75% protein and has a well-balanced profile with no gaps in the amino acid profile. Due to the lack of anti-nutritional factors, restricting amino acid profiles, or digestibility restrictions, it is especially well-suited for feeding aquaculture or poultry. Certain co-products are produced by various fermentation processes, and each has the potential to grow in the future.

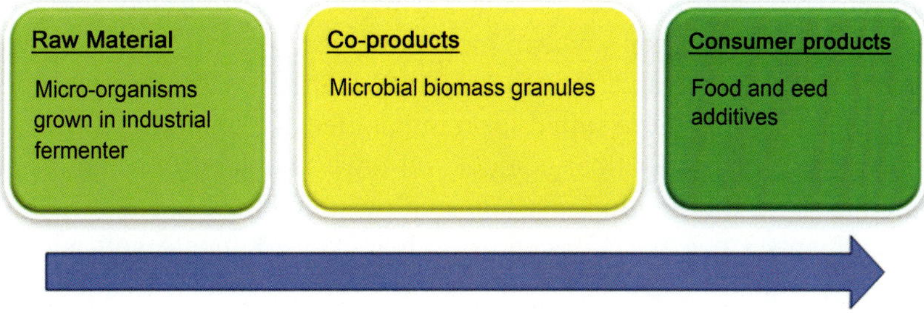


## 4.3.7. Oilseed Crushing and Vegetable Oil Refining

High-value food items such as oils from soybeans, rapeseed, sunflower seeds, and linseed are also being used more and more to make biodiesel. The food that's left behind after the oil is extracted has a lot of protein. For farm animals that produce food, oilseed meals are the primary source of proteinaceous feed ingredients. Historically, seeds were mechanically crushed to extract oil (a process known as "expelling"), which produced cakes. Since crushing produces a larger percentage of oil than protein meal separation, it has replaced protein meal separation as the most often used method. In addition to lowering the possibility of biological contamination, toasting is employed to lessen the presence of any anti-nutritional elements that can impair the digestibility of protein meals. It is possible to argue that soybeans' high protein content is now the primary reason for their cultivation, although the oil from soybeans, rapeseed, sunflower seeds, and linseed has the highest commercial value. Due to market forces, soybean meal has taken the lead globally as the primary source of protein for the feed industry, establishing the standard for all other vegetable protein sources. Because of soybean meal's exceptional qualities—which include a very high score for digestibility, palatability, protein concentration, amino acid profile, and nutrient density—it has established itself as the industry standard.

Additionally, feed purchasers and animal nutritionists choose soybean meal first because of its consistency, affordability, and all-around availability, which includes the opportunity to use price-hedging techniques. This is especially true for feeds intended for pigs and poultry. Vegetable protein meal is commonly used in food pellets, as protein is a crucial component in feed formulation. Because of its exceptional versatility, soybean meal can be fed to a wide variety of farm animals. Research and development on soybean meal has produced a wide range of adaptable products for various uses in animal nutrition, like soybean protein concentrate, which contains more than 70% protein and is especially good for aquafeed and young animals with immature digestive systems. Another illustration is extruded full-fat soybean meal, which is great for chicken feeds and has less protein but more easily digested energy. The most popular substitute for soybean meal is rapeseed meal, which is also the principal source of vegetable protein produced in the EU. Similar to sunflower meal, rapeseed meal is more suited for ruminant feeds than monogastric or fish feeds because to its higher fiber content and lower availability of amino acids compared to soya. Rumination-resistant sunflower and linseed cakes have been fed to ruminants since the 17th century. Their lack of innate anti-nutritional components, which means they don't require further special treatment to be used as feed, is one of their advantageous characteristics. Because sunflower cake contains a lot of the amino acid methionine, it's a great diet for birds that lay eggs.

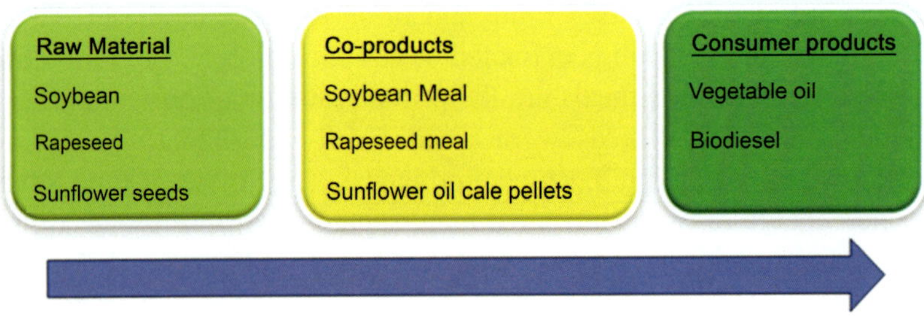


## 4.3.8. Gelatine Production

Gelatine has a long history; during the Pharaonic era, gelatin-like substances were made in Egypt. Gelatine-prepared fish and fruit specialties were regarded as a delicacy and presented at feasts, according to written documents. In the pharmaceutical and cosmetic industries as well as in contemporary cuisine, gelatine is a necessity. The raw materials used to make edible gelatine come from healthy animals that have been killed in an abattoir and whose corpses have been determined to be suitable for ingestion after antemortem and postmortem examinations. Highly valued feed components are among the byproducts of edible gelatin manufacturing. These byproducts can be safely used in feed for both food-producing and non-food-producing animals, fully complying with the Animal By-Products Regulations (EC) No. 1069/2009 and 142/2011. Dicalcium phosphate dihydrate, a byproduct of the gelatin manufacturing process, is an important source of calcium and phosphorus for farm animals used in food production. Before calcium phosphate is precipitated, crushed bones are degreased and demineralized with diluted hydrochloric acid. The final product can be used to make aquafeed, poultry, pigs, and pet feed. Because processed bone phosphates are more digestible than rock phosphates and minimize phosphate excretion into the environment, using them as a phosphorus source in chicken diets has been demonstrated through scientific studies to support sustainable animal husbandry practices. Large volumes of animal lipids are also produced during the production of edible gelatine. Like other nutrients like protein, fiber, carbohydrate, and sugar, fat is necessary for livestock diets to maintain optimal performance. Moreover, it enhances the flavor of feed and serves as a valuable source of energy. Dried animal proteins that come from the synthesis of edible gelatine derived from raw materials in accordance with Regulation (EC) No. 853/2004 are known as gelatine process derived proteins. Highly digestible amino acids including lysine, valine, arginine, and leucine that are appropriate for animal diets can be found in proteins produced via the gelatine process.

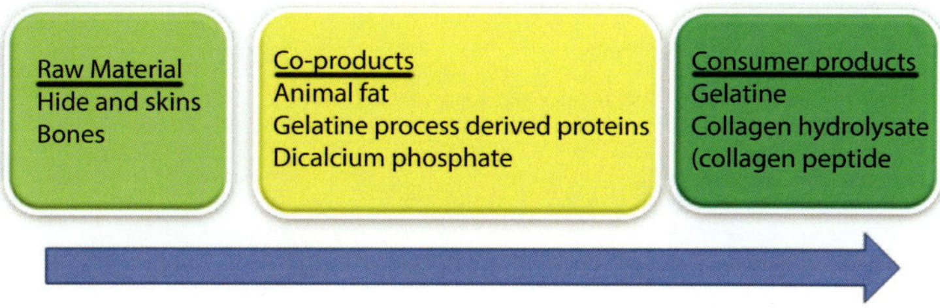


## 4.3.9. Dairy Processing

The majority of people are aware that milk is used to make cheese, but fewer are aware that whey, a liquid byproduct, is left over after the coagulation process caused by the addition of rennet or the decrease of pH. Different types of whey, such as acid whey and sweet whey, are created depending on the process and the animals from whom the milk was obtained. Whey products are mostly composed of lactose, protein, and minerals rather than much fat, which is utilized to make cheese. Different products are produced from the processing of liquid whey, which can be used in animal feed or as vitamins and sports drinks in human meals. These latter consist of fat-filled whey, whey protein concentrate, whey powder, and whey permeate; the lactose-to-protein ratios of these products are usually variable. Given the scale and growth of the cheese industry, whey powder is a practical answer from a logistical standpoint. The most popular application for whey-derived products is as a milk substitute feed ingredient for young ruminants, like calves and lambs. These are extremely tasty and easily absorbed dairy-based feed additives that improve animal performance and digestive health while increasing appetite and feed intake. Whey is reasonably rich in calcium, phosphate, sodium, potassium, and chloride, and it has a great amino acid profile. There are no anti-nutrients in it. For young ruminants, skimmed milk—typically in powder form—is another well-known milk substitute feed ingredient. However, due to its use in human food products, its use has been progressively declining in Europe over the past 10 years. Skim milk is defined as milk that has had the majority of its fat removed in order to make butter; all of the protein is still present, but there is very little fat. This is very digested and has a high biological value. Although the majority of the fat-soluble vitamins (A and D) are lost along with the fat, skim milk is an excellent source of water-soluble vitamins. The quantity of milk replacers given to young ruminants, such as veal calves, has gradually decreased in favor of concentrated feed and roughage in recent years, while skimmed milk powder has progressively been replaced by alternate vegetable feed materials.

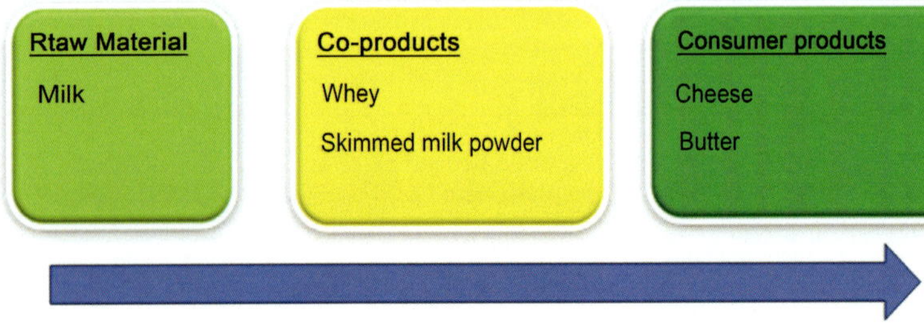


## 4.3.10. Meat Production/Animal By-Products

Animals provide high-value byproducts like leather and gelatine in addition to meat. One can also prepare a variety of animal byproducts not intended for human food that is not intended for use by humans. A variety of byproducts are converted into valuable resources for the feed business by the animal rendering sector. All of the resulting feed materials are derived from animal materials that are processed in compliance with the guidelines set forth by the Animal By-Products Regulations (EC) No. 1069/2009 and 142/2011. These animal materials are categorized as "fit for human consumption at the point of slaughter" (also known as "category 3 animal by-products" in the EU). Animal products are sterilized and stabilized throughout the rendering process with the application of heat and pressure, allowing for storage and subsequent processing. Animal fat and processed animal protein (PAP) are the two major products of rendering. Animal byproducts are also the source of a number of specialty items utilized in the feed sector. Regulation (EC) No. 999/2001 also specifies the usage of these feed additives, including limitations for specific species. For several reasons, feed producers especially benefit from rendered goods. First of all, they provide easily absorbed nutrients like minerals, fats, and proteins. Furthermore, because European laws are the strictest globally, feed ingredients are always of the highest caliber and safety. Additionally, feed manufacturers can count on a steady supply of ingredients made in Europe. Environmental advantages also exist: these materials have a minimal carbon footprint since the primary product (meat and dairy) bears the majority of the environmental burden rather than the animal byproduct.

Since 2013, the use of processed animal protein (PAP) derived from non-ruminant animals in aquaculture has been authorized. PAP is especially well-suited for feeding carnivorous fish, including salmon. PAP includes lipids, minerals, such as calcium and phosphorus, and important amino acids, such as methionine and lysine. In addition, it contains no anti-nutritive ingredients and is very tasty and digestible for fish. As of this writing, there is talk about relaxing the feed ban that was first implemented in 2001. The use of pig PAP in pig diets and poultry PAP in poultry diets is the next potential relaxation. The efficacy of controls based on analytical tests to confirm the identity of specific types of PAP is a necessary prerequisite to this. An essential element in the feed and pet food industries is animal fat. It is the most often used kind of fat in the feed business and is valuable to feed manufacturers due to its physical and nutritional qualities. You can utilize animal fat in animal feed for any kind of animal. The only hard fat produced in Europe is tallow. Another significant animal by-product is liquid poultry oil, which has a linolenic concentration similar to that of rapeseed oil. Blood plasma that has been spray-dried is a crucial component of young animals' diets, particularly for weaning piglets. Its immunoglobin content strengthens health performance and increases feed conversion ratios while assisting the piglets' growing immune systems.

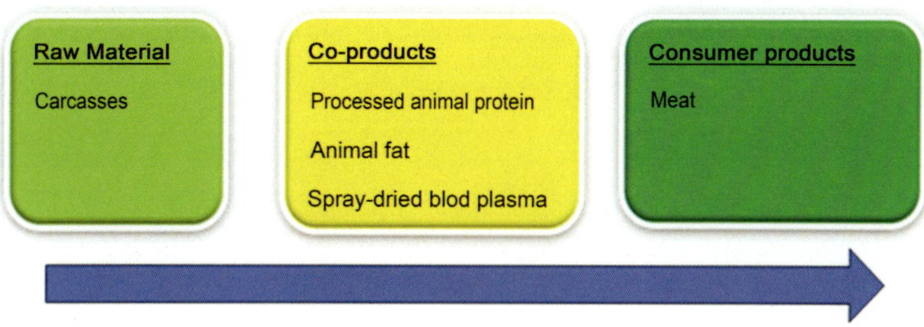


## 4.3.11. Food Manufacturing

In the process of making goods like bread and biscuits. Certain products, such as pasta, chocolate, cereal for breakfast, and crisps, usually don't satisfy the manufacturer's specifications and end up as leftovers meant for animal feed. Production faults can cause food to be broken or wrongly shaped, colored, flavored, or branded, failing to fulfil the required requirements. Surpluses of unsaleable goods can also happen after a product line is ended or after a seasonal event like Christmas or Easter. The difficulties in delivering some foods on a daily basis might also lead to previous foods. Any leftover inventory is usually still appropriate and safe for use in the making of animal feed after it is no longer commercially viable for human consumption and the manufacturer or merchant has thought about donating it to a charitable organization (like a food bank). It is possible to gather various foods and transform them into an energy-dense diet that can replace components that are often chosen for their energy content (such as wheat, barley, or maize). When compared to cereal grains, some previous food products, such as chocolates, crisps, and croissants, may even be deemed "fat-fortified" because of their oil content, providing additional nutritional benefits. Previous foods are easier to digest since the starch and other nutrients have usually also been heated. In addition, mycotoxins do not harm properly maintained historical foods. Former foodstuffs are a very broad category of feed materials; however, since they are intended for animals that will produce food, they cannot include any fish or meat. It's also critical to clarify that they are not the same as catering trash because they come from food manufacturers and retailers. All types of agricultural animals receive processed leftover meals as feed, but pig feed is the most common usage for them. Although giving chocolate to farm animals may worry some people, happily, they metabolize theobromine—a substance that can be lethal to dogs—in the same manner as people. Compared to previous instances in this publication, former foods stand out since they are not co-products in the conventional sense. Nonetheless, they show that the feed business can provide answers and make the most of the resources at its disposal at every point in the food supply chain.

## 4.3.12. Other Co-Products

- **Grape Seeds and Pulp:** Co-product of the pressing of grapes for making wine or grape juice. A source of polyunsaturated fatty acids and beneficial antioxidants.
- **Cotton Seed Meal:** Co-product of oil extraction from cotton seeds. Cotton seed meal is a good source of protein, however lower amino acid availability and the presence of anti-nutrient gossypol.
- **Glycerine:** Co-product from the production of fatty acids and esterification. An energy-rich feed material that is particularly suitable for dairy cows.

- **Palm Kernel Meal:** Co-product of the palm kernel oil extraction process. A high-fiber medium-grade protein feed, most suited to ruminants.
- **Olive Oil Cake:** Co-product from olive oil extraction. After being defatted and destoned, frequently used as pig feed in areas local to where olive oil is produced.
- **Groundnut Meal:** Co-product from peanut oil extraction. Has a high protein and oil content.
- **Feed Beer:** Beer that is not destined for human consumption.
- **Broken Eggs:** Broken eggs from hatcheries and egg packing stations are typically used to produce egg powder.
- **Fish Trimmings:** These include the food-grade parts left over after the fish has been filleted, along with the frames (heads and backbones) and the viscera. Collectively, they make up about 54% of the fish meal used in Europe (2016, IFFO).
- **Linseed Meal:** Co-product from linseed oil extraction. The meal is protein-rich and contains omega-3 fatty acids.
- **Apple Pomace:** The solid residue that remains after milling and pressing of apples for cider, apple juice or puree production.
- **Malt Residuals/Culms:** Co-product from the malting industry. Provides a good source of protein, energy, and fiber.
- **Wheat Middlings:** Consist of the byproducts from the wheat milling process that are not flour (e.g., screenings, bran, germ, and flour remnants). They are richer in digestible fiber and protein compared to flour.
- **Copra Meal:** Co-product from coconut oil production. Rich in fiber with mid-range protein levels.
- **Oat Hulls:** Co-product of oat rice or oat flake production. Very high fiber content, low in protein and energy. Widely used in the diet of some ruminants.
- **Hydrolyzed Feather Meal:** A protein-rich co-product from poultry processing operations (broiler, turkey, and others). For food producing animals in the, it is only allowed in aquafeed.

## 4.4. SHIFTING FROM A BY-PRODUCT TO A CO-PRODUCT MENTALITY

The distinction between these two phrases is not legally defined, but FEFAC has long pushed for suppliers to adopt a co-product mentality rather than a by-product one. Generally speaking, a by-product is an inevitable residue that needs to be disposed of cheaply. Co-products are given the same consideration and care as the primary product because of their significance as nutritional resources. Using a co-product effectively goes beyond only adhering to safety regulations in accordance with the law; it frequently leads to changes in the primary processing methods that guarantee a co-product of higher quality. It is supported by an integrated value chain that is intended to facilitate this and acknowledges that maintaining the nutritional integrity of the resource has additional socioeconomic worth.

Chapter 5

# Animal Feed Processing

## Contents

## 5.1. OVERVIEW

The application of physical, chemical, biochemical, biological, physiochemical, and engineering techniques to feeds and fodders in animal systems to improve nutrient utilization is known as animal feed processing. The following are the goals of feed processing:

- To make the feed more palatable;
- To detoxify or remove undesirable ingredients;
- To make the storage easy and safe;
- To increase nutrient content and nutrient availability;
- To change the particle size or density of the feed;
- To make animal production more economical.

To attain the best conversion rate, an animal or bird's diet must be balanced, and obtaining complete or compounded feeds for various groups of animals or birds involves a few processes. It shouldn't be limited to feed formulation alone since improved results from balanced feed formulations shouldn't be guaranteed. It is always necessary that the farm or feed mill has gone through multiple screening stages before properly processing (compounding or milling) an entire feed. Feed made at home, on your own farm, or at a feed mill should always be of higher quality, fresher, and less expensive. Feed processing is the process of applying various treatments to feedstuffs in order to improve their nutritional quality. While some therapies are straightforward and simple to administer, others call for sophisticated and time-consuming procedures. Do you know why processing feed is necessary? The following benefits of feed processing:

- Improves starch availability and protein utilization;
- Eliminates or inactivates the anti-nutritional factors in feed ingredients;
- Ensures preservation of nutrients in feed; and
- Improves appearance and palatability of feed.

## 5.2. PROCESSING TECHNIQUES

The most widely used feed-processing methods are discussed in subsections.

### 5.2.1. Chopping Green Fodder and Hay

Do you believe it is beneficial to chaff green fodder? Indeed. Before giving the animals any green fodder or hay, it is imperative to cut it. This facilitates easier digestion and keeps the animals from selecting certain parts of the feed, such as the leaves, and leaving the stems behind. Is it necessary to chop every kind of fodder? No, is the response. Depending on the kind of fodder and the kind and class of animal it must be fed to, cutting the fodder may be necessary. Generally speaking, fine-stemmed green fodder such as lucerne, grass, berseem, and so on, doesn't need to be chafed. Nevertheless, chaffing works best with green fodder with thick stems, such as maize and jowar. Because the animal is likely to leave back the thick stems if they are not chaffed. Before preserving green fodders like jowar and maize as silage, they must also be chaff-treated. Hay chopping yields very little benefit for sheep, cattle, and horses. Additionally, chopping reduces waste and costs significantly, especially when dealing with hard-stemmed and coarse-textured hay.

### 5.2.2. Grinding

Feed is ground to a smaller particle size, which facilitates the rapid and efficient action of the digestive enzymes. The age of the animal or birds and the seed's size, shape, and hardness all affect the particle's size. Because the outer covering of seeds and grains is difficult for animals, especially the very young and old, to digest, grinding is done. While sheep and goats chew quite effectively, they might not need to have their grains well-ground for cattle, pigs, or horses.

### 5.2.3. Mixing

Ingredients for feed are often combined in a mixer to create a diet that is nutritionally balanced. During farm mixing, the ground ingredients are layered on the pucca floor, with the largest ingredient at the bottom and the remaining ingredients layered on top of it according to their proportion. On top are the premix and mineral supplements. After that, the components are combined by hand or with a shovel until a homogenous consistency is achieved. For optimal mixing, the lot should be small and require at least three to five turns. These days, considerably less expensive combined mill and grinding or mixing equipment that may generate 100–500 kg of feed per hour are available; these units are specifically made for farm mixers.

### 5.2.4. Pelleting

These days, pelleting is a popular technique that makes use of concentrates, hay, or mixed diets. In addition to being economical, it also boosts feed intake, reduces feed waste, and frees up storage space. The steps involved in making pellets are as follows: first, food is heated to between 65 and 80°C with a moisture content of roughly 17% to soften feeds; next, the softened material is pressed through a suitable-sized mold to form pellets; finally, the pellets are cooled by coolers that are positioned beneath the pelleting machine, and finally, crumble is produced in a crumbling machine. For the pellets or crumbles to endure handling in large quantities and not shatter readily, they should be hard and stiff.

## 5.2.5. Roughage Processing Methods

Roughages' physical properties can be changed in a number of ways to increase intake and/or digestibility. These consist of pelleting, cutting, and grinding. Four categories can be applied to these techniques:

- **Particle Size:** Decreasing the size of the particles to expose more of the material to pre-feeding treatments or to create more surface area for microbial fermentation of fibrous components in the rumen or hind intestine.
- **Handling:** To create a substance that is more manageable or compacted for ensiling or storing.
- **Density:** In order to make the substance denser so that animals can consume more.
- **Mixing:** To combine other components with the roughage in order to improve the animal's capacity to digest the fibrous material and/or absorb more of it, as well as to balance the nutrients offered to the animal.

Based on whether water content is added to or subtracted from the roughage, there are two types of roughage processing:

- **Dry Processing Method:** These techniques lower the water content to the appropriate degree. This covers dehydration as well as grinding, chopping, bailing, pelleting, and cubing.
- **Wet Processing Method:** In order to optimize feed intake and roughage digestibility, soaking is the act of mixing or spraying water on roughages to soften their stems and ensure consistent mixing of concentrates with roughage. This comprises Juicing and Soaking in Green

### *5.2.5.1. Grinding to Reduce Particle Size*

Before being pelleted, roughages are typically ground, with pellet sizes ranging from 12/64" to 48/64" (4.8 mm to 19.1 mm). In comparison to long hay, which weighs 5–6 lb/cft, pelleted roughages weigh approximately 40 lb/cft. The consumption of roughage significantly increases when low-quality roughages are pelleted. Feed intake appears to be optimized when a concentrate mixture is added at a 30% level during the pelleting of whole feeds; without this addition, feed intake decreases. However, ruminants fed pellets with a higher concentrate level may develop parakeratosis, a condition where the rumen papilla deteriorates. In other words, pelleting diets deficient in forage has negative effects.

Roughage must be ground before pelleting and mixing. These mechanical processes enhance voluntary intake, improve nutritional content, and facilitate the preparation of complete meals. For roughage fed on its own, the material should be ground to a length of 1–2" (2.5–5 cm). When included in complete rations, roughage should be ground to 0.5–1.0" (1.3–2.5 cm). Adding 1% tallow or water during the grinding process can help reduce dust loss. Adding molasses to ground hay makes it highly palatable and increases feed consumption.

#### *5.2.5.1.1. Impact of Grinding Roughages*

- Rumination and rumen retention time are decreased when ground roughage is fed to cattle. An increase in feed intake results in improved animal performance.
- Because feed passes through roughage more quickly when it is finely ground, crude fiber digestibility is typically decreased. Dairy cows fed finely ground grain produce less rumen acetate, which lowers the amount of butter fat in the milk.

Given the expense of the machinery, the rising energy costs associated with operating the machinery, and the transportation of straw from the farm to the feed plant and back, it seems that this approach is not viable at the farm level and may even be dubious at the commercial scale. When low-quality roughage is ground, it becomes more digestible of dry matter than when high-quality roughage is. Before feeding, straws and stuvers should be coarsely ground or chopped. Despite being one of the most basic mechanical procedures for treating any feedstock, this work is still costly and not very pleasant. Expensive mechanization is needed to handle the vast amounts of roughage efficiently, and grinding consumes a lot of energy. Hay or straw grinding is projected to cost more than $20 per ton, even in large, efficient plants.

### *5.2.5.2. Chopping*

Another name for it is chaffing. The forages are finely or coarsely cut into little pieces. By avoiding selective feeding, chopping minimizes the amount of plant material wasted. The device known as the chaff cutter is employed for the specified purpose. Because chopping increases bulk density and exposes a sizable portion of roughage surface area to microbial digestion, it both makes handling easier and enhances digestion.

### 5.2.5.3. Baling

It's one of the most widely utilized techniques for improving fodder handling convenience. After being chopped, the fodder is left to dry on the field. Forage that has been dried is baled using a field or stationary baler. In wealthy nations, it is extremely well-liked. Using this technique, the feed is gathered and spread out across the field to dry. Hay conditioners can be used to help dry out fodder in highly humid environments. At the time of baling, the fodder should have a moisture level of between 15 and 20%. Roughages are bundled into handy-sized bundles called bales, which are typically shaped like squares or circles. Round bales may be stored in the field without causing much damage because they are an efficient way to shed rainwater. Bales range in size from 20 to 120 kg.

### 5.2.5.4. Cubing

Wafer manufacture is modified by cubing. Long hay has a density of 7 pounds per cubic foot, whereas cubed hay and pelleted hay have densities of 25–32 and 40 pounds per cubic foot, respectively. The hay is broken rather than ground when the cubs are born. Thus, the cube contains the fewest possible fine particles. The majority of the cubing process uses premium lucerne hay. Water is sprayed on the hay during cubing to bring the moisture content up to roughly 14%. It is challenging to form a cube that is satisfactory if the hay contains more than 10% grass. There are two types of cubers: stationary and movable. It is well-liked in industrialized nations.

### 5.2.5.5. Pelleting

Before pelleting, roughages are typically ground, with pellet sizes ranging from 12/64" to 48/64" (4.8 mm to 19.1 mm). Pelleted roughages weigh around 40 lb/cft, compared to long hay, which weighs 5–6 lb/cft. The consumption of roughage increases significantly when low-quality roughages are pelleted. Feed intake is optimized when a concentrate mixture is added at a 30% level during pelleting; otherwise, intake decreases. However, ruminants fed pellets with a higher concentrate level may develop parakeratosis, a deterioration of the rumen papilla, indicating that pelleting diets low in forage has a negative impact.

### 5.2.5.6. Dehydration

Alfalfa and lucerne are examples of green fodder that can be maintained by briefly (three to five minutes) drying them at a high temperature (600–1,500°F) in a dehydrator. Usually, high-quality, young, growing fodder is used. There is no leaf loss during the forage preservation process, and the highest possible level of protein and dry matter is retained. Carotene is lost (5–15%) during the industrial drying process. Typically, cattle are fed dehydrated lucerne pellets (17% CP) as a supplement rather than as their main source of roughage. Comparing these pellets to cubed or baled hay, they are hardly appetizing.

### *5.2.5.7. Soaking*

In order to optimize feed intake and roughage digestibility, soaking is the act of mixing or spraying water on roughages to soften their stems and ensure consistent mixing of concentrates with roughage. It is not necessary to soak green roughages when they are chaffed; they can be fed straight away or combined with concentrate mixture or dry roughage.

### *5.2.5.8. Chemical Treatment of Roughages*

Worldwide, enormous amounts of fibrous crop leftovers, including rice, wheat, and barley straws, are underutilized. The chemical treatment of straws and other low-quality roughages to enhance their quality for ruminant feeding attracted a lot of attention in the 1970s and 1980s. Alkali treatment is the main chemical technique used to improve the digestion of roughages. This extracts the indigestible lignin from the digestible carbohydrates in cell walls by breaking (hydrolyzing) chemical connections. Early studies that treated roughages with sodium hydroxide demonstrated highly exciting increases in their digestibility. In addition to the anticipated gains linked to the roughage's greater digestibility, sheep and cattle fed roughages treated with alkali demonstrated benefits in the form of increased feed intake. Because of this early success, other alkali sources such as potassium hydroxide and calcium hydroxide were looked at. Additionally, research was done on the application of ammonia to roughages, specifically with regard to treating straw with urea, anhydrous ammonia, or urine as a substitute source of ammonia. Although applying hydroxide to low-quality roughages is a very effective technique to increase digestibility, intake, and enhance animal performance, the application of chemical treatment under commercial conditions has been limited due to several logistical, safety, and environmental concerns. The foundation of the wood pulping process used to make paper is alkali treatment. The alkali has the effect of cleaving cellulose, lignin, non-starch polysaccharides (NSP), and internal connections. Enzymes that hydrolyze materials can more easily access cellulose thanks to the degradation of lignin and NSP. In the process of treating fibrous material with alkali, the structural NSPs are partially soluble as well. Heat treatment combined with an alkaline environment can dissolve lignin and produce free phenols. Because lignin melts at temperatures above 160°C and auto hydrolysis occurs, steaming at these temperatures can improve digestibility.

### *5.2.5.9. Sodium Hydroxide Treatment to Increase Digestibility and Intake of Roughages*

Scientists had demonstrated that boiling straws in a solution of sodium hydroxide (NaOH) might increase their feeding value before the 1920s. Dr. Roy Kellaway and associates from Sydney University in Australia created workable methods for handling massive amounts of grain stubbles when baling. The use of NaOH has three main drawbacks. The first is the potential for operators to be harmed by the strong NaOH solution and splashed. It is extremely caustic and especially harmful if skin or eyes are exposed to it. The excessive salt content and its detrimental effects on the environment are the second problem. Excessive salt intake can also be harmful to the animal; kidney damage has been seen in dairy cattle given high sodium diets over extended periods of time. Excessive salt concentrations can also be detrimental to the ecosystem and the structure of soil. One further issue that is present in all alkali treatment procedures is equipment corrosion and damage. Ammonium hydroxide is a weaker alkaline agent than NaOH, thus it generally results in smaller gains in roughage digestibility. Ammonia has a preservation action that prevents the growth of mold during treatment, and it increases the amount of rumen degradable N that the rumen microorganisms may use to produce protein. These benefits offset the negative effects of ammonia.

### *5.2.5.10. Treatment with Urea or Ammonia*

If urea is introduced to plant biomass in the natural environment, it will be quickly broken down into carbon dioxide and ammonia due to the urease activity of many common bacteria found on plant materials. As soon as there is enough water, ammonium hydroxide is produced.

Urea → Ammonia + Water → Ammonium hydroxide

Applying an aqueous solution of urea to straw is safer than using ammonium hydroxide. Numerous experiments have been conducted to find the ideal urea concentration and the ideal interval between the application and feeding of the straw. It is commonly accepted that before using straw, it should be covered with plastic sheeting or ensiled for at least 4 weeks. Approximately 5% urea (50 kg urea per ton of straw) should be utilized. The treated straw is stable for an extended amount of time after it is placed within an airtight pit or container. Ammonia solutions can be substituted with ammonia gas, often known as anhydrous ammonia. About 3% of anhydrous ammonia is needed to achieve the same result. Anhydrous ammonia is still widely used in some regions of Europe where vast stacks of cereal straw are wrapped in black plastic before gaseous ammonia is introduced from a mobile tank. This method has the benefit of being simple to apply to large stacks of straw as a gas. The process could be hazardous because the gas is combustible and poisonous. The success and appeal of this process are largely dependent on how the roughage is harvested, chopped, and treated. Other important factors include having adequate storage facilities and equipment for feeding out the treated material. It is ideal to have highly mechanized ways for harvesting and filling the pit or silo because it can involve massive amounts of material.

### *5.2.5.11. Biological Treatment of Roughages*

This may include:

- Enzyme treatment: Cellulase solution is sprayed on straw at 25 mg/100 kg straw.
- Fermentation: Chopped straw is pretreated with 3–5% NaOH and steamed at 120°C for 15 min; then fermented with bran-type media cultured with cellulolytic microorganisms at 40–50°C for 2 days.
- White-rot fungi, mushrooms, and other microbes.

The metabolic barrier brought on by lignin, which occurs in a range of 3 to 13% of lignocellulosic straws, limits the effective utilization of these straws. While some white-rot fungi, such as *Phanerochaete chrysosporium*, break down lignin to a degree of 65–75%, other fungi, such as *Coriolus versicolour* and *Ganoderma applanatum*, break down more than 45% of the lignin present in lignocellulosic materials. Species that only break down lignin and not hemicelluloses are preferred. The Indo-Dutch Project on Bioconversion of Crop Residues has studied the white-rot Basidiomycetes, which are edible and frequently non-toxic mushrooms.

- **Zadrazil Process:** *Pleurotus* sp. was used to treat the straw. There are significant losses of organic stuff in the process. It is inappropriate for farmer-level small-scale activities.
- **Karnal Process:** Fundamentally, it is a non-sterile biological treatment of ligno cellulosics in a solid substrate fermentation (SSF) system that results in a potential improvement in the increase of straw quality. This is a two-step process that involves pretreating cereal straws with 4% urea and 40% moisture, then ensileing the material for 30 days. In the second stage, the urea-treated material is thoroughly mixed with 0.1% calcium oxide and 1% single superphosphate. It is then moistened to 60–65% before being inoculated with 3% *Coprinus fimetarius* (an alkali tolerant strain) culture grown on millets. When *C. fimetarius* reached the mycelial stage of development, the solid substrate fermentation was stopped.

There are numerous benefits of using urea in the initial period. In addition to cleaving the ligno-carbohydrate bonds in the treated straw, ammonia also functions as a chemical sterilant to stop the growth of undesirable organisms, raises the CP content from 3–4% to 12–14%, and helps create a favorable environment (high pH). In the second step, the fungus produces amino acids while ensnaring more free ammonia in the straw that has been treated with urea. As a result, the amount of amino acids in the fungal-treated straw increased significantly. There were significant losses of dry materials. However, by implementing certain adjustments, dry matter losses were decreased from 35% to 7%.

### *5.2.5.12. Conservation of Roughages*

Roughages and crop residue are agroclimatically and seasonally variable. As a result, processing, shipping, and storage are required for the extra green fodder produced during the season. The primary goal of conservation is to preserve fodder with little nutrient loss. In order to ensure maximum nutrient availability, fodder is chopped before it reaches maturity. There are three methods for preserving fodder: airtight chambers, artificial driers, and sun drying (Hay and Silage).

1. **Straw:** It has the highest percentage of crude fiber and maybe the lowest protein level of all the foods. They are high in silica but rather low in phosphorus, accessible calcium, and trace elements. Since wheat or rice is our staple grain, wheat or paddy is the cereal crop that is most frequently grown. Straw is the leftover material from a crop that has had its grain harvested. Farmers gather these straws from the fields and store them in large stock for animal food after they have dried. Usually, it is kept elevated above the ground on a platform made of bamboo or wood. To reduce spoiling in locations with high rainfall, this is necessary. Straw is the leftover stems and leaves from cereal crops after the grains have been separated. For instance, straw from paddy, wheat, barley, oats, and so on.
2. **Stover:** After the grains are separated, the stems and leaves of millet crops are included in this category. Usually harder than straws, the stover must be chopped into smaller pieces before feeding, for example, Bajra, Jowar, and the stover of maize.
3. **Hay:** It is the term for cut dried grass or clover that is used as animal feed. Hay's composition is influenced by the climate and grass availability. For making hay, the best plants to use are rye grass, Italian rye grass, coastal Bermuda, orchard grass and other natural

species; legumes include alfalfa and clovers (red and white); plant materials include oat, barley, and wheat. Animals are typically given hay, especially during the winter or when pasture isn't accessible due to drought or other circumstances. Hay's nutritional content is determined by the crop's maturity level when it is harvested for hay production. Even though the latter cuttings give a higher weight, the earlier cuttings are more nutritious. When in flower, high-quality herbage is trimmed early in the morning and left on the field. Farmers attempt to harvest hay at the moment when all leaves are fully formed and seed or flower heads are just short of full maturity in order to maximize the nutrient value of the crop. The chopped fodder is turned after four to five hours, and by the afternoon of the next day, the moisture level has dropped to 25%. Hay is sent for cutting after drying. Hay is often gathered into bales or bundles and then moved to a central area for storage once it has been cut, dried, and raked into windrows. Moisture content should not exceed 23% for safe storage, which can be achieved by rotating the hay frequently and distributing it over a large area.

4. **Silage:** Making silage is the process of fermenting chaffed cereal green feed in an anaerobic environment. The process involves gathering a forage crop with a high moisture content (over 50%) and then fermenting it in a trench, pit, tower, bunker, or plastic silos. Green fodder can be preserved for up to a year or six months if proper care is given because microorganisms help convert the sugar in the feed into lactic acid when it is in an anaerobic environment (no air inside the pit or tank). Cereal green fodder, such as maize, sorghum, bajra, hybrid Napier, sugarcane tops, oats, and so on, is preferred for making the highest quality silage because their higher sugar content allows microorganisms to employ fermentation to produce lactic acid. Furthermore, the stiff stems of these cereal fodder crops require longer drying times. Squeezing good silage releases moisture and has a unique acidic, fruity scent. Its color is yellow-brown. Corn and jowar are the best crops to use to make silage. Other crops that can be used to make silage include non-leguminous fodders (with the exception of turnip), hybrid Napier grass, Sudan grass, berseem, oats, cowpea, and millets.

### *5.2.5.13. Hay Making*

Hay is the most common way of storing extra fodder for usage, which is dried and stacked. The process of drying cut grass or crop to lower its moisture content from 70–90% to 20–23% is called hay making. To preserve as much nutrients as possible, it is ideal to harvest the crop or grass before it flowers; nevertheless, in the majority of grain harvests, the stovers are dried and turned into hay once the grains reach maturity. Despite making up a large portion of the planet, pastures, grazing grounds, and grasslands rarely have the quantity or quality of feed available year-round to support increased livestock production. To attain the intended animal production in such a scenario, supplementing forages or crop leftover, such as hay from cultivated land, is inevitable. Furthermore, the bulk of animals are kept in mixed farming systems, and in these circumstances, crop leftovers play a significant role. Hay can be traditionally made from grasslands, but it can also be made from fodder crops that are grown. Thus, addressing the fodder shortfall during the lean periods is the main goal of hay production. Foraging for usage during times of scarcity has been done since ancient times. The practice of utilizing crop waste and straws for animal feed dates back almost as far as cereal production. Although the process of generating hay appears to be ancient, it is the most significant method of preserving pastures and feed, even though it depends on favorable weather throughout the crop-harvesting season. One of the old-fashioned methods of conserving forage and agricultural wastes for use during lean and scarce periods, as well as storing them in hay banks for use in times of natural disaster, is the production of hay.

### *5.2.5.14. Hay Types*

Roughage with more than 18% crude fiber and less than 20% crude protein is referred to as hay. It usually has to be used in conjunction with other foods and is never a complete meal. Straws, stovers, and crop wastes are valuable sources of lean-season nutrition for ruminants and are frequently combined with hay. Hay can be categorized according to the crop it is made from and the degree of technology employed. According to technology, the primary varieties of hay are:

- **Long Hay:** This kind of hay is traditionally made by tying the crop into little bundles, turning them, and storing them in a stack. This is the most typical method of producing hay.
- **Chopped Hay:** This kind of hay is made in areas with ideal drying conditions and highly automated systems. It is more suitable for mechanical handling and less voluminous.
- **Baled Hay:** It is first produced manually by hand and subsequently by stationary machinery. The most common and easiest to construct are round bales. Compared to conventional bales, their shape better sheds rain and resists water.
- **Hand-Trussed Hay:** It is commonly used in hand-crafted Hay, frequently to lessen breaking. Hay that has been wafered and

pelleted is easy to handle, store, and transport since it is dense and free flowing.

- **Dried Grass:** Grass and crop leftovers are artificially dried at high temperatures in this approach. This kind is not financially viable.
- **Barn-Dried Hay:** Here, raw materials are air dried with a fan to prepare hay. It's rare to find this kind of hay prepared.

Depending on the crop from which hay is made, they may be classified as follows:

- **Legume Hay:** Leguminous plants, such as lucerne and various clovers, are used to make them. In general, they include higher levels of proteins, vitamins, and minerals than non-leguminous foods. Young calves are typically given this hay in calf-starter meals.
- **Non-Legume Hay:** Less proteins, minerals, and vitamins are present in them. This kind of hay can be made from grasses or crops. Carbohydrate content is often high in those manufactured from green crops like barley or oats.
- **Mixed Hay:** A mixture of non-leguminous and leguminous plants is used to make them. Since they do not reach the flowering stage at the same time, such a crop is typically harvested early. The amount of non-leguminous and leguminous plants in the hay will determine its nutritional value.

Compared to crops with thick, stiff stems, those with thin, flexible stems are more practical and appropriate for creating hay. Oats have a narrow leaf-to-stem ratio and a thin, hollow stem, which makes them the ideal crop for creating hay. Other legumes ideal for creating hay include range legumes like stylosanthes and siratro, as well as cultivated fodder crops like lucerne, cowpea, sun-hemp, and berseem. Legume hay is more nutrient-dense and is typically offered to animals in addition to other fibrous feeds like stover and straws. As long as they are collected when their development is at its peak, both annual and perennial crops—mostly grasses—make excellent hay. In India, the majority of hay is made from crops including millets, sorghum, and maize, which are often gathered following grain harvesting. Additionally, pre-maturity grass from protected grasslands is collected, dried, and stored as hay. The hay making is influenced by the following two primary factors:

- **Season:** When excellent weather is predicted, pasture (either naturally occurring or seeded) can be managed by grazing or by taking silage cuts as needed to ensure that the herbage reaches an appropriate stage for curing. In order to maximize the use of labor and equipment, a farm that plans to produce a large amount of hay should think about cultivating a variety of species. This will extend the season during which the various crops will be at their best for hay making.

- **Stage of Crop Maturity:** The crop's maturity level is one of the main variables affecting the quality of the finished product when creating hay. As a raw material reaches maturity—especially after flowering—its quality rapidly declines. But compared to more established herbage, extremely young, green grass is more difficult to manage. In the early stages of crop growth, fodder has high protein and calorie contents, but yield is low. Thus, one must strike a balance between quantity and quality; typically, pruning between the first and middle stages of flowering will result in a respectable harvest. Lower nutritional digestibility is the result of diminished protein and other nutrient levels as the crop reaches maturity beyond its maximum potential. For hay-making purposes, the best time to harvest a crop is between the third and half of the bloom stage in grasses, the grain milk stage in cereals, and the sensitive pod development period in legumes. Furthermore, crops grown in Kharif may be harvested in early September, when the rainy season ends, while crops grown in Rabi may be harvested in February or March when the temperature begins to rise. Younger material can be cured more successfully in "good" Hay-making climates than in damp climates or in places where tropical cyclones are a possibility. Compared to pasture grasses, cereals used for hay are cut at a more mature stage, frequently when the grains have fully developed.

Farmers may still produce hay in areas where there is a general lack of fodder, even if the crop is of low quality. For instance, a lot of hay is made from over-mature herbage in tropical and subtropical regions because the weather is unfavorable for drying during the season when the grass is at its optimum. Although the product is subpar, hay—which is just as valuable as straw—is nonetheless greatly prized.

### *5.2.5.15. Haymaking Process*

The following basic steps are involved in all methods of haymaking:

- **Selection and Cultivation:** Choose appropriate plant species and cultivate them in healthy soil using suitable techniques such as fertilization and watering. A high seed rate is necessary to achieve a dense crop stand, which produces finer stems, larger yields, and fewer weeds.

- **Cutting:** Lucerne should be cut at the appropriate maturity level. The first cut is made at the early bloom stage, while the second and subsequent cuts are done when 10% to 25% of the crop is in bloom, or when new shoots begin emerging from the crown. For grasses, cutting is complete when they begin to head out. Cowpea crops are cut when pods are halfway to fully developed, and grain crops are cut at the milk stage.
- **Curing:** Proper curing ensures that hay retains as many nutrients as possible while reducing moisture levels enough to prevent mold growth and spoilage.
- **Preventing Shattering:** Minimize the loss of leaves during curing, especially in leguminous plants, which are more prone to shattering. Extra precautions are necessary to avoid this.
- **Avoiding Losses from Bleaching and Fermentation:** Losses from bleaching and fermentation occur when the curing temperature is too high or when moisture content is not properly managed.
- **Preventing Leaching:** Nutrient losses from rain-induced leaching are the primary concern during the monsoon season. Leaching losses are minimal if rain occurs soon after cutting, but increase as the curing process progresses.
- **Moisture Content:** The target moisture content for hay is 25% to 28% for loose hay, 22% to 25% for baled hay, and 19% to 22% for chopped hay.
- **Storage:** Hay should be properly protected from sunlight and rain to maintain quality.

The common methods of haymaking are field curing, mow curing, and artificial drying.

## *5.2.5.16. Properties of Good Hay*

- When fodder is cut at the right time, it has the most nutrients and produces superior hay. If fodder is not harvested for haymaking on time, the stems will become woody and lignified, which will lower their nutritional value. Therefore, only properly staged crops can produce high-quality hay.
- Proper hay will have a significant amount of the recommended leaves. Hay that has been broken throughout the haymaking process will consequently be of lower quality.
- Hay's green hue is a sign of its quality. It indicates that losses from

bleaching, leaching, fermentation, and other processes have been kept to a minimum and that a larger percentage of the leaves have been retained.

- For it to be smooth and appetizing, thin-stemmed crops must be used.
- Weeds and toxic plants should not be present in the crop that is used to make hay.
- When stored, hay should have no more than 20–23% moisture content. More moisture will cause fermentation, which will produce a lot of heat. In the end, it can result in spontaneous combustion. It would have significantly impacted the nutritional value even if it did not catch fire.
- Dust and mold will not be present in good hay.
- Its palatability will be increased by its nice scent.

Mostly, hay is fed in combination with some green forage. Additionally, ammonization is done to enhance and improve its digestibility. Intake and digestibility are enhanced by chafing. Hay is sometimes supplied straight off the stack in our nation's arid regions, which leads to significant waste. For increased animal productivity, chaffing, and mixing with green feed or fodder are therefore necessary. If stored properly, hay can be preserved for extended lengths of time after it is prepared. Hay is stored with the intention of keeping it dry and shielding it from being wasted by rodents, insects, wandering animals, fire, or wind. Hay should ideally be stored near its intended use rather than in the field itself. The conventional way of storing a lot of hay is in loose material stacks. Hay and straw are frequently stored in tree forks, with straw being stored in particular to keep it out of the reach of roving animals. In semi-arid regions, house-top (roof) storage is highly prevalent in smallholdings. Flat-roofed homes can shade the livestock from the sun and act as platforms for hay and straw storage.

### *5.2.5.17. Silage Making*

Making silage is another crucial way to preserve green grain. Silage is the fermented byproduct of green forages, where the sugars are preserved by the acids created during anaerobic fermentation. The process involves chaffing green fodder and keeping it in specially built subterranean or above-ground structures, making sure that no moisture or air gets inside. Thus, conserved green fodder ferments in an anaerobic environment without significantly reducing the nutritional content of the material retained. Preservatives are created by the acids that are produced during the fermentation process. 'Silage' is the term used to describe the resultant product. Silage offers a number of benefits.

- The entire forage plant cannot be maintained in a succulent state other than by the ensiling technique.
- When the crops are at their most nutrient-dense in their development, they can be harvested and stored.
- Higher amounts of nutrients are retained in silage than in hay because less is lost to shattering, leaching, and bleaching.
- Meal energy is preserved to 85% by silage. Hay only keeps 80% under ideal circumstances and 50%–60% under unfavorable ones.
- Stem crops that are too thick to be used to make hay can be used to make silage.
- You can make silage in nearly any type of weather.
- Silage crops yield more than other types of hay crops.
- Compared to hay, silage requires less storage space.
- Compared to hay and other dry crops, silage poses a lower fire risk.
- Silage has a mild laxative effect and is easier to eat and digest.
- Since cobs, shucks, and coarser stems are consumed, there is no waste in feeding.
- A greater source of protein and carotene is silage.
- In addition to the primary fodder crop, weeds can also be used.
- Ensiled fodder's nutritional value may be raised by adding the right supplement.

### *5.2.5.18. Crops Used for Ensiling*

The majority of fodder crops can be used to make silage. Since they seem to lose these traits while ensiling, several crops that are unsatisfactory as hay or green fodder due to bitterness or terrible flavor are good as silage. Both thick-stalked and weedy crops can be successfully ensiled. Which crop is most frequently used to make silage, can you guess? Because it produces a large yield, maize is a preferred crop for silage. It also has the necessary dry matter percentage and accessible sugars at the cutting stage for ensiling. Consequently, natural fermentation is guaranteed without the need for preservatives. Another significant crop for silage is sorghum. Grain sorghum is not as good for silage as sweet sorghum. Hay-crop silages are the term typically used to describe silages made from grasses and legumes. A good silage should taste acidic, have a hint of green to it, and have a subtle, pleasant smell. It should have enough acid to stop further deterioration and be devoid of sliminess and mold growth. Ensilage can be successful when the following prerequisites are met:

- Make the appropriate crop choice. Enumerating all the crops that can be utilized to make silage is going to be challenging. All grasses, lucerne, cowpea, oats, barley, red clover, and so on, are among the crucial ones.
- Choose fodder species, such as sorghum and maize, which have a high natural sugar content.
- Steer clear of legumes with high protein and low sugar content, as this promotes the growth of harmful microorganisms. If not, prepare mixed silage by blending lucerne with corn, sorghum, and berseern, and so on.
- Because they contain low amounts of sugar and large levels of protein, immature crops should not be cut for silage.
- Because the sugar content of crops grows throughout daylight hours, they may be cut late in the day.
- Before ensiling, add molasses as evenly as possible to crops that are low in sugar.
- When crops are drenched from rain or dew, don't cut them. Before moving the crop to the silo, let it wilt in the field if needed.
- Because sugars can produce water through oxidation, quickly consolidate and shut the silo.
- Stop rainwater from seeping into the silos.

### *5.2.5.19. Types of Silos*

There are a great many distinct kinds of silos in use. These range from enormous concrete buildings outfitted with all the tools to basic plastic bags.

- Concrete tower silos have a diameter of 8 to 10 meters and a height of 15 meters. When the silage is ready for use, it is withdrawn through a series of apertures on one side and replaced with fresh fodder that is blown in at the top.
- As the name suggests, trench silos are horizontal, shallow trenches. For fewer animals, the trench's length can be 2–3 m (A); for more animals, it can be longer (B). Because of its easy accessibility, labor expenditures are minimal. The ideal soil for trench silos is clay with a low water table. To improve accessibility, the trench can frequently be carved into the slope of a hill.

- Pit silos are precisely that—small silo pits sunk into the ground at an elevation above the surrounding area, to produce silage.
- Tub silos are upright, easily hand-unloadable silos with a very large diameter.
- Concrete or brick bunker silos are ideally suited for locations with loose soil, high water tables, and a fear of caving in. These are constructions above ground made of walls that have compressed ensilage between them. Concrete or wood may be used for the walls. In a similar manner, feed is added to and removed from trench silos and bunkers. Tractor trolleys, trucks, or carts filled with chaffed feed can be driven through, unloaded, and into. Then, blade-equipped tractors are filled with fodder.
- Where the soil is suitable and the water table is high, clamp silos can be employed. To raise the walls' height, a shallow trench is dug, and soil is placed on both sides. Pack ensilage as densely and as high as you can. On top, a high-quality plastic seal is fitted.
- Silage is made in a variety of methods using plastic sheets. Plastic film can be used to prepare silos of different diameters.
- Silage bags are composed of tiny plastic tubes. These are 50 kg or less tiny silos. Thus, low-cost silage structures are available for small livestock operations. Each tube's ends are doubly knotted to create a bag with handles on both sides.

The amount of fodder to be ensiled and the viability of the project determine the silo's size and shape. Typically, silos in villages are made of pits and bunkers. Nonetheless, towers or buns are typical silos at government and commercial dairy farms where huge amounts of silage are to be produced. The silos might be rectangular or cylindrical in design, and they can be pacca or kaccha. Plastic bag silos are used in dairying in the peri-urban area. They have little losses and are easy to operate.

### *5.2.5.20. Silage Making Process*

The following steps are involved in silage making:

- To achieve a moisture level of 65–75% (25–35% DM), the fodder crops intended for silage production should be harvested or allowed to wilt in the field.
- Harvested fodder may be cut or ensiled as such. However, the silage crops are often diced into 2.5–5 cm pieces. Chopping makes it easier to handle silage after it has been packaged for fodder.

- Inside the trench or silo, the chopped fodder is tightly packed and shut off from the outside environment.
- In the ditch or pit, make sure the chopped fodder is distributed evenly. To eliminate as much air as possible, press (trample) the chopped material with your feet, a tractor, or bullocks, depending on the size of the pit. Fill the silo's top or the pit's uppermost level with feed up to one meter (four to six feet) in height.
- It is necessary to consistently press (trample) every layer of chaffed hay that is 1/2 meter thick.
- Cover the feed from all sides with a layer of karbi, wheat busha, or paddy straw, about 14 to 15 cm thick. To seal the material and stop air and water from getting in, apply mud first, then plaster. Verify sure there are no gaps in the silo.
- As an alternative, the packed fodder can alternatively be covered with a plastic sheet. Mud plastering is one way to seal the sheet's edge.
- Watch it from time to time and repair any cracks or holes right away. Silage would be available in 45 days.
- To lessen fermentation losses, different commercial preparations, cane molasses, limestone, urea, and ground shelled maize are occasionally added. To increase the nitrogen content and palatability of cereals and grasses, urea and salt are added at 0.5 and 1%, respectively. Molasses is added to grass silage at a rate of 3.0–3.5% to increase the sugar content.
- Keep the remaining silage covered while you open the silo pit from one end only and remove the necessary amount.

Animals may not like the taste of silage for a few days at first. For the first week, add 5–10 kg of silage to their green fodder ration to help them develop the taste. About 20–34 kg of silage and other fodders may be fed per head each day after the taste is established. The silo pit must be filled quickly, making sure that each step—chaffing, packing, sealing, and so on—is completed. Only an airtight silo can guarantee higher quality. It must be opened from one side and promptly closed when opening for feeding.

### 5.2.6. Processing of Grains

It is convenient to categorize grain processing techniques into dry and wet procedures. The main goal is to increase starch's availability and digestibility, as it makes up between 70 and 80% of grains. Nevertheless, the process of doing this is difficult because:

- the type of starch varies among grains in its digestibility; and
- availability of starch even varies from one grain variety to another, particularly in milo.

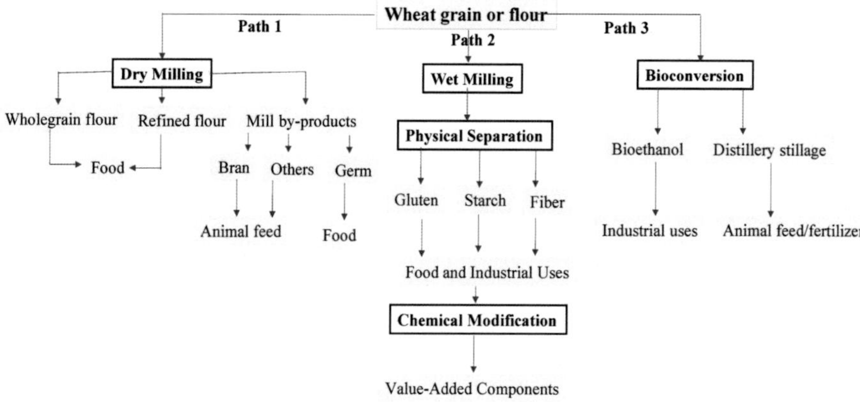


Source: Abdel-Aal, E.-S. M. (2024). Insights into grain milling and fractionation practices for improved food sustainability with emphasis on wheat and peas. Foods, 13(10), 1532. https://doi.org/10.3390/foods13101532.

The most powerful enzyme systems against hydrated starch are those found in animals and microorganisms; nonetheless, it is uncommon for animal feed to degrade grains to the point where all of the starch is hydrated. Nonetheless, it does happen during the steam-flaking, pelleting, and bursting of grains to some degree. For most animals, breaking open the pericarp to reveal the endosperm is the minimal amount of processing needed to guarantee effective grain digestion. The degree of rolling and grinding, as well as the amount of particle size reduction, are factors in the second processing stage. The amount of surface area exposed to microorganisms and digesting enzymes, as well as the number of starch granules liberated from the endosperm's protein and non-starch carbohydrate matrix, are all influenced by particle size. Individual starch granules that are securely bonded inside the endosperm matrix and shielded from enzymatic digestion can be found in even the smallest particles. The third stage of processing deals with the scenario where starch granules are firmly anchored within the endosperm matrix. In order to expose the starch to enzyme digestion by hydration and/or gelatinization, it may be necessary to utilize high temperatures with or without water to rupture the granules. Whole grains break down by a combination of soaking and grinding in the crop and gizzard, so they are easily fed to sheep, who can chew them efficiently, and to chickens. To enable complete development and effective breakdown of grains, the gizzard must adapt. While the gizzard's adaptation period is not an issue for layers, feeding whole grains to broilers can lower feed conversion efficiency (FCE) as there is less time for the establishment of an efficient crop and gizzard function. It is imperative to shatter the seed coat of cereal grains, particularly the smaller ones, using mechanical or chemical treatments before feeding cattle and pigs since their capacity to chew them is limited. One exception is oat grain, which horses and cattle can utilize effectively even when fed whole and unprocessed. The main methods of processing grains, their effects on the grain, and the implications for animal digestion are compiled below:

1. **Grinding:** The process of reducing a feedstock to a particle size by impact, shearing, or attrition is known as grinding. Apart from soaking, this is the most popular, affordable, and straightforward method. There is a large range of equipment available, and all of them permit some degree of particle size control. While finer ground grains are more prevalent in poultry and swine, coarser ground grains are preferred in ruminants.
2. **Dry Rolling or Cracking:** The process involves moving grain between a tightly spaced set of steel rollers—which are typically surface-grooved—without the use of steam. It produces an end product of coarsely ground grain, sometimes called flaking, and splits the hull and/or seed coat. Cattle are often better off when fed flaked grain rather than finely ground grain.
3. **Popping:** Most people have heard of popped rice and popped maize, which are made by quickly applying dry heat to grains, which causes the grains to suddenly expand and burst the endosperm. This approach can be used to increase the digestibility of any grain, but it seems to work best when processing sorghum or other milo grains. Popped milo's light density means that it needs additional storage space.
4. **Extrusion:** Grain extrusion requires a lot of moisture, pressure, and heat. When the feed being extruded passes through a barrel and runs into flow resistance, frictional heat is produced. Steam can be injected into the barrel, or it might be steam jacketed. Extrusion occurs at temperatures between 125 and 170°C, although only for a brief period of time (15 to 30 seconds). Getting a high degree of starch gelatinization is the main goal of extrusion. In terms of the combination of fermentation and digestive processes in ruminant animals, the relationship between the degree of gelatinization and the physical properties of the final feed is not well documented. However, the effect of extrusion cooking on the digestibility of various cereals is well understood for human and monogastric nutrition.

5. **Micronizing:** The grain goes through three stages in this process: soaking, rolling, and high-temperature infrared burners. In theory, the process is comparable to steam flaking, leaving the grain mostly intact but denser and more vulnerable to amylolytic digestion.

6. **Roasting:** The process of roasting involves heating the grain to the correct temperature over a period of time in an oven or passing it over a flame. This causes the grain to expand somewhat and yields a product that is edible. The process can be applied to whole soybeans to eliminate heat-labile inhibitors and enhance the nutritional value of pigs and poultry.

7. **Pelleting:** It is a popular commercial process in which moisture, heat, pressure, and a mechanical process combine to form a larger particle from smaller ones. The heat and steam utilized throughout the conditioning process (typically 10–15 seconds) as well as the heat produced by friction as the feed goes through the dye partially gelatinize starch. Another benefit of pelleting is that it gives you control over the density and end pellet size, as well as flexibility in terms of the feed's particle size to be pelleted. Pellets can be formed into tiny pieces or cylinders with varying lengths, diameters, and hardness levels. Thus, it provides a way to regulate the place and pace of digestion. The cost of the procedure is approximately 10% more than that of non-pelleted concentrates.

8. **Dehulling (Corticated Form):** Dehulling is the process of removing the outer coat from grains, nuts, and some fruits because in pigs, poultry, and other monogastric animals, the hulls are high in fiber and low in digestibility. The barley, oat, and rice hulls are the most well-known outer layers of cereal grains. Hulls are now marketed as by-products after being mixed with other milling leftovers from various cereal grains. Such unhulled (undecorticated) oilseeds like soybeans, cottonseeds, groundnuts, sunflowers, and safflowers have comparatively low protein content.

9. **Microwave Treatment:** While it is well recognized as a quick and energy-efficient cooking method for food preparation, its efficacy in processing grains for animal feeding has not received as much attention. This may be helpful now that continuous microwave processing is possible, but it may have been hampered by batch processing associated with older microwave technologies.

10. **Soaking:** Livestock feeders (which are not mechanically processed) have long used hard grains soaked in water for 12 to 24 hours to feed horses and mules with sore mouths. Oilseed by-products, such

as mustard cake, can also be beneficially soaked in water to reduce toxicity factors like HCN.

11. **Steam Rolling:** Grain is compacted into flat pieces through the process of rolling, which involves moving the grain between rollers. Another name for steam rolling is "crimping," which is the process of rolling grain after it has been exposed to steam for a brief amount of time—typically one to eight minutes. Compared to dry rolling, the product made by steaming the kernel appears more intact and crimped. Compared to dry rolling or grinding, steam rolling provides little to no benefit in terms of feed efficiency. Nonetheless, very young animals whose teeth have not yet fully grown or very old animals whose teeth are badly worn down may find the product helpful.

12. **Steam Flaking:** The whole grain is heated with steam for 10 to 40 minutes during this procedure, after which it is rolled to different degrees. While the entire grain stays intact, the process causes the endosperm and seed coat to separate, increasing the surface area of the grain. Furthermore, a large portion of the starch is gelatinized during cooking, increasing its vulnerability to amylase attack. The degree of intestinal digestion is significantly influenced by the amount of cooking. The grain is rolled to create a rather flat flake after being exposed to high moisture steam for long enough to increase the water content to 18–20%. Thin flakes work better because they facilitate the more effective rupture of starch granules, resulting in a more appealing texture.

13. **Pressure Cooking:** This product is a lot like flaking grain that has been steam-processed. When pressure flaking occurs, the grain is exposed to steam under pressure for a brief period of time—one to two minutes at 50 psi (pounds per square inch). Steam is introduced into the cooker until the temperature of the grain within the chamber approaches 300°F. After being removed from the cooker at a temperature lower than 200°F, the grain is flaked with 20% moisture. The flakes made by pressure flaking are less brittle than those made by steam flaking.

14. **Exploding:** Grain explodes when it steams under pressure and then releases the steam into the air. To reach 250 psi, high-tensile strength steel "bottles" are filled with steam. A valve opens after roughly 20 seconds, allowing the grain to escape as swollen balls without hulls. The kernels absorb moisture under high pressure and swell to many times their original size when released into the air.

15. **Reconstitution:** Reconstituted grain is mature grain that has been harvested at the typical moisture level of 10–14%. Water is then added to raise the moisture level to 25–30%, and the wet product is kept for 15–21 days in an upright silo (for necessary compaction) before feeding. When the grains are removed, they are rolled and ground. When it comes to fattening cattle, properly reconstituted sorghum and steam-processed flakes milo yield comparable outcomes.
16. **Ensiling:** Grain can be partially converted from starch to organic acids, primarily lactic acid, through ensiling, which helps to preserve the grain. Microbial enzymes can target the starch granule more easily after ensiling. Silage's high moisture content promotes endogenous enzyme activity up until the point at which fermentation causes the pH to drop. Silage can be kept in storage for a long time as long as anaerobic conditions are kept.
17. **Chemical Treatment:** Protein microbial breakdown is decreased when feeds are treated with hydroxides, and the resistance of the seed coat to digestion is decreased when formaldehyde is used. For some feed components, formaldehyde has been used as a microbial barrier to protect protein. Formaldehyde-treated protein can be used to coat starch, fat, or other substances to preserve the entire complex from microbial attack. Given that different protein fractions are soluble in alcohol or alkali, these substances may be applied before endosperm pretreatment, especially in the case of sorghum grain.

### *5.2.6.1. Effects of Storage and Processing Grains*

The following are the primary methods by which processing and storage can modify the chemical and physical characteristics of grain:

- Particle size reduction;
- Moisture addition;
- Heat treatment (affect CHO and protein);
- Addition of chemicals

#### *5.2.6.1.1. Particle Size Reduction*

It employs roller mills and hammer mills to reduce the size of the particles. Grain grinding fineness is measured using one of two techniques. The particle is described using geometric mean diameter and standard deviation in one way. An alternative approach depicts the average and variance as fineness and uniformity modules, respectively. A number of variables, including grain type, moisture content, temperature, screen size, flow rate, tolerance setting, and roller type, affect how finely ground something gets.

### 5.2.6.1.2. Moisture Addition

Wet feed has been shown to increase growth rate and feed efficiency in pigs by about 5%. It has been demonstrated that reconstitution, or three weeks of anaerobic storage, improves cattle digestibility, especially of the protein fractions. Sorghum reconstitution results in a drop in glutelin concentration and an increase in alcohol soluble fraction, which should improve digestibility but lower protein quality. Tannins are inactivated by moisture, which is another advantageous effect.

### 5.2.6.1.3. Addition of Chemicals

Alkali makes the exterior starch particles swell, which breaks down the seed coverings and partially gelatinizes the outside starch granules. This increases the digestion of fiber in fibrous grains. NaOH has a higher solubility than Ca (OH)2, making it a more effective preservation. Ammonia effectively inhibits the growth of mold, but it requires an atmosphere devoid of ammonia. With the exception of a decrease in the amount of α-tocopherol, organic acids have minimal effect on the physical and chemical characteristics of grains when they are preserved by fungi like acetic acid and propionic acid.

## 5.3. LIVESTOCK FEED PROCESSING EQUIPMENT

One sector where compromises with subpar animal feed are strictly prohibited is the agricultural industry. One of the most crucial pieces of equipment used in the business to provide animals with the proper amount and quality of nutrients for healthy growth is livestock feed processing equipment. Therefore, while thinking about a livestock farm for cattle, poultry, and other farm animals, investments must be made in the right appropriate equipment for effective feed production. Equipment used to process feed for livestock usually refers to devices that turn raw materials into feed. To achieve consistency in delivery in terms of quality and efficiency in distributed feed, the equipment includes livestock feed crushers, mixers, pellet makers, coolers, and packers. These machines streamline the entire manufacturing process. Livestock farmers and feed makers can benefit greatly from investing in equipment for preparing animal feed. Here are a few specific advantages:

- **Improved Feed Quality:** The processing of feed ingredients to comply with nutritional values and the formulation's homogeneity are ensured by the use of specialized equipment for animal feed processing. Therefore, enhanced nutrient distribution throughout feeds and consequently higher-quality feed may also result from greater evenness. A higher milk or meat yield can be attained via improved development, better reproduction, and overall health, all of which are correlated with better feed quality. Pelletizing feed increases its digestibility. Because pelletizing would make the feed easier for the animals to absorb and digest, the feed's entire nutritional value is nearly certain to reach the animals. The nutrients that are contained are preserved throughout feed processing with no loss at all thanks to contemporary technology.

- **Increased Productivity:** Because automated feed processors process feeds so quickly, they might process them considerably faster without the need for manual labor, which lowers the possibility of human error. In large-scale operations when productivity and time are critical, this would be required. Via a single, streamlined, integrated line, mixing services, grinding, and pelleting operations can all be automated. Maintaining animal nutrition requires consistent feed in both quality and quantity, which is provided by machinery.
- **Cost Savings on Livestock Feed Processing Equipment:** On the other hand, while this may only be the one-time cost of the livestock feed processing equipment, long-term savings can add up to quite a bit. Effective feed processing minimizes waste, maximizes the utilization of raw materials, and lowers operating expenses.
- **Stable Production with Livestock Feed Processing Equipment:** The equipment used in the preparation of livestock feed will provide the best possible quality, consistency, and dependability in the final product. This will be crucial since maintaining the animals' health at optimal levels will require such constancy. A consistent supply of feed generally reduces the likelihood of disruptions and unfavorable effects on the farm's productivity and animal health. The equipment is simply programmable to maintain consistent feed quality in accordance with the various nutritional needs of the various types of cattle.
- **Adequacy and Flex:** With today's equipment for manufacturing livestock feed, you might tailor each formula to the unique nutritional requirements of various livestock species. It provides a chance for variable diet planning, which is crucial for farms that look after a variety of animal species.

- **Better Biosecurity:** By ensuring that the feeds are processed under hygienic and controlled settings, advanced feed processing equipment lowers the risk of infection and improves biosecurity.
- **Environmental Benefits:** Modern machinery is made to use less energy, which guarantees strong output rates. Because of this efficient processing, the environmental impact of the feed manufacturing process is reduced.

Factors to consider when buying livestock feed processing equipment include:

- **Cost and Price:** Compare machine cost and price difference to stay within budget without compromising quality.
- **Supplier and Manufacturer:** Find a good reliable manufacturer or supplier of high-quality, durable, and dependable equipment.
- **Specifications of Equipment:** Consider specifications like capacity, power consumption, and compatibilities with the rest of your machinery in the line of production.
- **Maintenance and Support:** Ensure that there is good after-sales service and maintenance of the equipment as a fair way of maintaining your equipment for proper functioning.

## 5.3.1. Feed Milling Machines

### *5.3.1.1. Grinding Machinery*

One of the main tasks in the production of feed is grinding, or particle-size reduction. For many reasons, most feed mills run all entering ingredients through a grinder.

- Clumps and large fragments are reduced in size;
- Some moisture is removed due to aeration; and
- Additives such as antioxidants may be blended.

Each of these enhances the components' storability and ease of handling. Prior to further processing, substances in formula feeds are ground and subsequently sieved for additional purposes. In general, grinding components raise their bulk density and enhance their acceptability, mixing qualities, pellet-ability, and feed digestibility. Numerous mechanical and human processes including impact, attrition, and cutting are used to achieve it.

### *5.3.1.2. Hammermills*

The majority of hammermills are impact grinders that use steel bars that either swing or remain stationary to press items up against a solid, serrated portion called a striking plate or a circular screen. Until the material is as small as the openings in the screen, it is retained in the grinding chamber. Important factors that affect grinding capacity and product appearance include the number of hammers on a rotating shaft, their size, arrangement, sharpness, rotation speed, wear patterns, and clearance at the tip with respect to the screen or hitting plate. The amount of heat that the material receives from the grinding process depends on the air flow parameters and the length of time it is kept inside the chamber. Impact grinding works best with dry, low-fat substances, but it can also reduce the size of many other materials with the right screen choice and controlled intake. A "vertical" hammermill is more effective for specific substances, such as dry animal byproducts than a typical hammermill, which has a horizontal drive shaft that suspends vertical hammers. The hammers and screens are arranged horizontally in this mill, whereas the motor shaft is orientated vertically. After successfully reduced in size to the diameter of screen holes or less, the material is transported outside the mill by gravity, where it is stored in "make-up" bins or transported by air or conveyor. Unbreakable large particles fall through the mill and can be recycled or disposed of. As a result, foreign objects like metal and stones are released before they can damage the screen by forcing their way through (Figure 5.1).

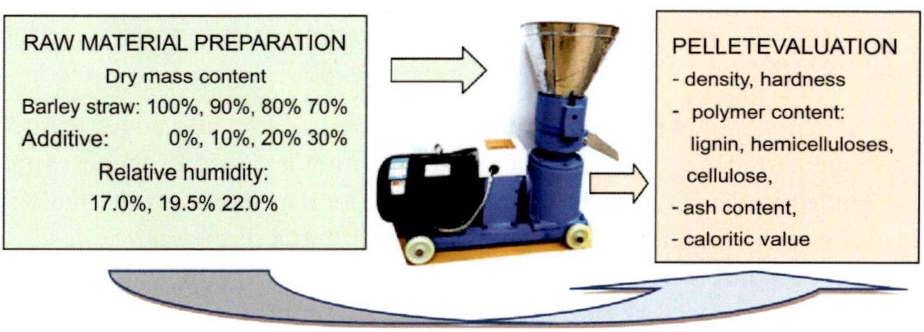


Figure 5.1. *An example feed pellet-making machine.*

*Source: Chojnacki, J., Zdanowicz, A., Ondruška, J., Šooš, Ľ., & Smuga-Kogut, M. (2021). The influence of apple, carrot, and red beet pomace content on the properties of pellet from barley straw. Energies, 14(2), 405. https://doi.org/10.3390/en14020405.*

### 5.3.1.3. Attrition Mills

Attrition mills partially apply the hammermill principle—that is, shattering by impact. They also impart a cutting and shearing action, though. Two discs with interchangeable worn surfaces are used for grinding. These discs can be turned, one or both of them; if both, their rotations are opposite. The assembly is used to shred and postpone when one disc rotates while the other remains fixed. In an attrition mill, materials that have been coarsely crushed by previous mills are frequently run through to blend or smooth out an ingredient or mixture that contains potentially clumpy liquids. An attrition mill's discs are often positioned vertically to allow materials that cannot be reduced to exit the grinding area via gravity.

### 5.3.1.4. Roller Mills

In roller mills, a combination of crushing, attrition, and cutting happens. These are rolls, either smooth or corrugated, that are spaced apart at a predetermined distance and rotate at the same speed, allowing material to pass between them. The addition of a tearing motion can be achieved by varying the roll rates and corrugations; for example, the bottom roll may have lateral corrugations while the top roll may have off-radial spiral corrugations. In order to create granules from hard pellets, the final type—known as a "Le Page cut"—offers a breaking surface with less impact that could lead to dust. Although roll grinding is cost-effective, it can only be applied to somewhat dry, low-fat materials.

### 5.3.1.5. Cutters

A type of grinder known as a rotary cutter primarily uses knife edges to shear dry particle particles against a hitting plate. Attrition and impact processes are also included in the mill, albeit their effectiveness is restricted if the material is easily reduced by cutting and the screen that limits discharge has a lot of holes in it. The mill is made up of a screen that takes up one-fourth of the 360° rotation and a revolving shaft to which four parallel blades are attached. It is ideal to use the mill to crack whole grains with as few "fines" as possible. It is not the last step in the process of making fish feed ingredients smaller.

### 5.3.1.6. Screening

In connection with the grinding of feeds, a sieving system that can sort materials into any desired particle size is needed. In this system, the "overs" can be rejected or re-ground. The "throughs" can be chosen based on the size preferences of the animal and combined in accordance with the formula's instructions. To remove the "dust" that is produced when dry feeds, particularly cereal grains, are hammermilled or impact ground, a dust-collecting device could be required. When organic waste sticks to the respiratory system and becomes a food source for bacteria or parasites, it can cause respiratory system disorders. This can happen when there is too much dust in the feed. Feeds going into the grinder can benefit from having a spray of oil or a semi-moist material like condensed soluble to help reduce the problem of excessive dust created during the grinding process. Mineral oil is sprayed into a hammermill chamber while the lucerne is being ground, resulting in a dust-free meal that resembles sifting crushed pellets.

### *5.3.1.7. Mixing Machines*

Starting with a certain combination of components known as a "formula" and adding up to a predetermined weight is the aim of feed mixing. This is prepared in such a way that every tiny amount of the total—a mouthful or a day's worth of feedings—has the same ratio as the original recipe. It is acknowledged that mixing is an empirical unit operation, meaning that mastery of the technique is primarily a matter of experience rather than science. Any and all combinations of solids and liquids may be used in feed mixing. There exist variations in the physical qualities of each constituent. Solids differ from one another in terms of their shape, size, density, electrostatic charge, and coefficient of friction as indicated by the angle of repose, elasticity, and resilience. Color, odor, and taste are also important factors to consider. There are variations in viscosity and density for liquids. "Mixed" can refer to either blended, suggesting homogeneity, or composed of disparate components, suggesting dispersion. Each of these definitions is combined when used to formula feeds; that is, mixing is the process of dispersing different components into a blend. It is unlikely, though, that particles in a sample will become uniformly distributed in terms of concentration or location. That is merely the aim of quality control. It has been proposed that "mixing and unmixing" would be a more appropriate term for a discussion on mixing because mixed particles always tend to separate during the procedure. The mixing process involves three mechanisms:

- The transfer of groups of adjacent particles from one location in the mass to another;
- Diffusion; distribution of particles over a freshly developed surface;
- Shear; slipping of particles between others in the mass.

Theoretically, randomness determines the position of particles inside a container, and random effects compound until they become more significant than the direct effects of mixing action. When liquids are mixed, random component movement produces homogeneity or order. Chance distribution leads to disorganization with dry solids. Disorder can be referred to as "random" when it has reached a roughly stable maximum. Particles in dry solids avoid an accidental or random arrangement due to a variety of causes. In actuality, combining feed ingredients may produce a specific pattern of non-random arrangement or particle segregation. Particle segregation results from variations in the physical parameters of the components as well as the mixer's shape and surface features. It's possible that the primary component producing segregation is particle size. Statistical techniques can quantify an improvement in mixing that gets closer to a random distribution of solids by reducing particle size. In general, during mixing, substances will get closer to a random distribution the smaller and more evenly sized they are prepared. To get an adequate quantity of particles of a vital ingredient (vitamin, mineral, or medication) for dispersion in every daily feed unit, a reduction in particle size is sometimes required in formulas. This might necessitate that the particle size be between 10 and 50 microns, the diameter of dust. When finely separated, some components become unstable and can pick up an electrical charge. When very small particles are mixed and these are much smaller than the mass of other ingredients, segregation can occur due to variables such as stickiness of oily and moist ingredients, roughness of the mixed material, and concentration of particles on a charged surface. Mixing can be done continuously or in batches. Batch mixing can be carried out using shovels on an open, level surface or in containers with fixed baffles or with moving augers, spirals, or paddles in the form of cylinders, half-cylinders, cones, or twin cones. The strategy of continuously mixing proportions by weight or volume works well for formula feeds that have few ingredients and little variation.

### *5.3.1.8. Horizontal Mixers*

1. Continuous Ribbon Mixers: The continuous or "twin-spiral" mixer is made out of a fixed, horizontal half-cylinder with rotating helical ribbons positioned on a central shaft. The ribbon and shaft spin inside the mixer to transfer materials from one end to the other. A few liters to several cubic meters can be considered capacity. Shaft rotation speed varies inversely with the outer ribbon's circumference, typically maximizing between 75 and 100 meters per minute. Any end may be utilized for discharge because the substance travels from one end to the other. You may clean these mixers by inverting them (Figure 5.2).

Figure 5.2. *Green Technik mod. PTM50, pelletizer used for the production of pellet samples in the small-scale company.*

*Source: Ilari, A., Foppa Pedretti, E., De Francesco, C., & Duca, D. (2021). Pellet production from residual biomass of greenery maintenance in a small-scale company to improve sustainability. Resources, 10(12), 122. https://doi.org/10.3390/resources10120122.*

2. **Non-Continuous Ribbon Mixers:** Different from continuous ribbon mixers, non-continuous or interrupted ribbon mixers include short sections termed "paddles" or "ploughs" arranged in a spiral around the mixer shaft. In contrast to continuous ribbon mixers, this type of action might be more effective when combining liquids and dry solids. These mixers come in an extensive range of sizes, with the outside diameter of the paddles travelling between 100 and 120 meters per minute.

### *5.3.1.9. Vertical Mixers*

A single or double screw (auger) positioned vertically through the center of a cylinder, cone, or hopper-shaped container can be found in a vertical mixer. Like a screw conveyor, the screw moves vertically to transfer incoming materials from the bottom (usually the intake) end to the top, where they are dispersed and fall due to gravity. The screw runs at speeds between 100 and 200 rpm. Until a blend is achieved, this process is repeated multiple times (typically within 10 to 12 minutes). It is also possible to load these mixers from the top. The findings indicate that vertical mixers are ineffective for evenly blending grains and liquids or for materials of quite different particle size or density. This unit is difficult to clean and there may be inter-batch contamination (Figure 5.3).

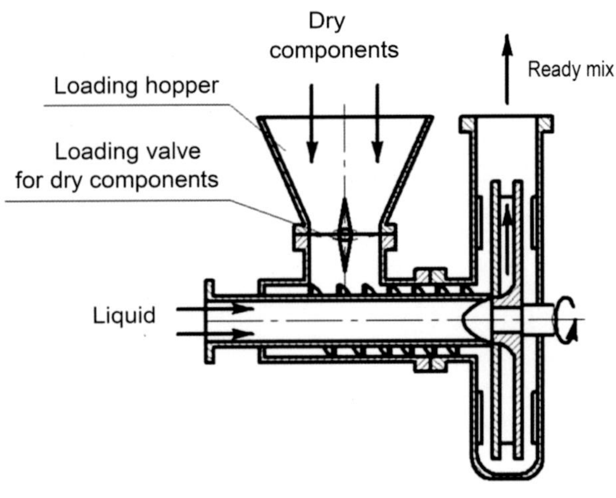


Figure 5.3. *Installation diagram for the preparation of liquid feed mixtures.*

*Source: Solonscikov, P., Barwicki, J., Savinyh, P., & Gaworski, M. (2021). Optimization of design parameters of experimental installation concerning preparation of liquid feed mixtures. Processes, 9(12), 2104. https://doi.org/10.3390/pr9122104.*

### *5.3.1.10. Other Types of Mixers*

The horizontal revolving drum is a third kind of mixer. This cylinder may have tapered ends on both ends and straight sides. The sides can be fixed with shelves or baffles to pick up and drop ingredients, or they can be smooth. This kind of mixer works best for blending materials that are dry, smooth, and have uniform physical qualities. A variant of this kind is the turbine mixer, which consists of a stationary cylinder with a rotating shaft that is attached to stationary paddles, plows, scrapers, or shelves intended for continuous material re-piling. This mixer is frequently used as a cooker to combine different kinds of fish meal into a standardized product and dry fish waste. Additionally, they work especially well for combining liquids into combinations that would clump or cake in a different kind of mixer as well as for mixing heavier materials. On soft materials like rice bran and lucerne leaves, some particle size reduction (grinding) may take place. Generally speaking, it takes 3 to 6 minutes to fully mix, unless there are lumps that need to be removed from the mixture due to the addition of liquid. A certain amount of centrifugal force is produced by the mixer shaft's-controlled rotation, but not too much. Originating in Holland, the "Nauta" mixer is built like an inverted cone with an inner rotating mixing screw that revolves around the interior wall. The mixer comes in a range of sizes, from very large production quantities to laboratory models for combining chemical and vitamin additions. It works incredibly well for adding modest amounts of liquids to dry ingredients and is great for premixing trace materials. The "entoleter" is a different kind of mixer that uses a high-speed spinning disc to forcefully throw the ingredient charge against a chamber's walls. This mixer works great for breaking up lumps or balls of compressed ingredients and will stop grain weevil eggs from hatching. However, it is not advised for all mixes because it may break vitamin A beadlets encased in gelatine.

### *5.3.1.11. Liquid Mixers*

Oils and preparations of water-miscible oils are frequently added to dry foods as a nutrient or energy source. The oil-soluble vitamins A, D, E, and K can be purchased in dry carrier concentrations, but the feed maker may also blend them in pure form. Nutrient-containing liquids can be combined more quickly and consistently than the same nutrient in dry concentrate conditions. As a result, the feed facility may require a liquid mixer. A horizontal tub or cylinder with several wires or paddles evenly distributed around an internal rotating shaft is the typical design of a liquid blender. Occasionally the shaft is hollow, creating a spray effect as liquids are driven through the paddles' perforations. While some types rotate at 1,200 rpm, others have a shaft speed of 400 to 600 rpm. Condensed fish or fermentation soluble, molasses, or oils are common premixed components that are blended with dry ingredients in a bowl-style variable speed mixer.

### 5.3.1.12. Mixing Operation and Evaluation

It is necessary to add components in a tested order from batch to batch in order to get accurate mixing. Large-volume components are often added first, followed by smaller-volume ones. Liquids should be added after mixing all dry components unless they are premixed beforehand. Total mixing time is important and is determined by the formula's composition. To prevent under- or overmixing, all mixers should be calibrated using laboratory recoveries of recognized ingredients (physically or chemically). Water soluble dye-coated, uniformly sized particles of iron, graphite, or salt are frequently employed as "tracers." For optimal results, each mixer should have its mixing duration and volume capacity calibrated.

### 5.3.1.13. Pelleting Machine

Three processes are used to turn a soft, frequently dusty feed into a hard pellet: adhesion, extrusion, and compression. A feed mixture is generally put through a conditioning chamber, where 4–6% water (sometimes in the form of steam) is added. In addition to providing lubrication for compression and extrusion, moisture also partially gelatinizes the raw starch that is present on the surface of vegetal ingredients when heated, resulting in adhesion. When feed enters the pellet mill, it changes from being air-dry (approximately 10–12% moisture) at room temperature to 15–16% moisture at 80–90°C in under 20 seconds. Friction causes a ring die's holes to expand and cause further compression and extrusion, raising the feed temperature to about 92°C. Pellets are air-cooled to just above room temperature in 10 minutes after being unloaded into a vertical screened hopper or into a screen belt of a horizontal tunnel drier. They are then dried to less than 13% moisture content. Finished pellets, contrary to popular assumption, include almost all of the nutrients found in compounded feedstuffs and additives. To meet formula needs, additional supplementation of thermolabile vitamins in the vitamin premix can make up for their loss, which can range from a little amount (as with vitamin C) to a significant amount. Although powdered enzymes introduced as an ingredient are inactivated, diastetastic enzymes (alpha and beta amylase) found in whole grains and cereal byproducts remain active even after processing by grinding and pelleting. Pelleting is a mechanical technique that includes pushing soft feed through apertures in a die made of metal in the shape of a ring. There are two types of holes: square or round, tapered or not tapered. A spinning shaft is turned on by one or more rolls that are positioned inside the die ring on a cam or eccentric when friction arises (because feed is present between the roll and the die). Feed is pushed incrementally through the die perforations, resulting in tightly packed layers of feed mixture when a finished pellet is dissected. The rolls only rotate when friction arises from the feed between the rolls and die, which is powered by a motor. In the conditioning chamber of the pellet mill or during formula mixing, water is frequently added to the dry feed to make the particles flexible for close compression, reduce friction, and absorb mechanical heat. Water can be offered in liquid or vapor form. The following is a possible list of variables arising from the equipment's mechanical operation:

- Pellet dies thickness as related to diameter of hole is a factor in pellet quality.
- Speed of rotation should also be considered for each die thickness/ hole diameter combination.
- The speed at which feed is introduced into the feed-conditioning chamber affects the moisture/temperature relationship which in turn relates to pellet quality.
- The amount of steam added to a given volume of feed should be in balance such that the drive motor on the pellet mill is operating at its maximum amperage. The opening of the steam valve at the pellet mill has a direct relation to the amount of water entering the feed as steam at any given steam pressure.
- Atmospheric conditions in the factory, especially relative humidity, which pre-condition the ingredients before processing, will affect die selection and operational settings.

Pelleting is more of an art than a science, as seen by the need to look at all of the aforementioned factors for every feed mix. It should be underlined once more that the pellet mill runs at its best when the motor amperage is used in accordance with the available voltage. Throughout pelleting, it's critical to constantly check the ammeter gauge (Figure 5.4).

**Figure 5.4.** *Food pellets.*

*Source: Chojnacki, J., Zdanowicz, A., Ondruška, J., Šooš, Ľ., & Smuga-Kogut, M. (2021). The influence of apple, carrot, and red beet pomace content on the properties of pellet from barley straw. Energies, 14(2), 405. https://doi.org/10.3390/en14020405 (Creative Common License).*

### 5.3.1.14. Extrusion

Feed extrusion is mainly reserved for high-value feeds, like aquafeeds, where the temperature during processing must be carefully regulated to prevent the degradation of heat-labile ingredients (e.g., enzymes, vitamins, and amino acids), the particle cohesion must be high to prevent water penetration, and the pellet density must be controlled (e.g., heavier for sinking versus lighter for floating feeds).

### 5.3.1.15. Drying and/or Cooling

The pellets are processed, screened to remove fines and broken pellets, and then sent to storage, where they are either bagged or loaded out in bulk. After cooling (horizontal or counterflow coolers are typically used) or dried and then cooled (horizontal equipment is used), the pellet temperature is lowered to ambient temperature (to avoid spoilage problems) (Figure 5.5).

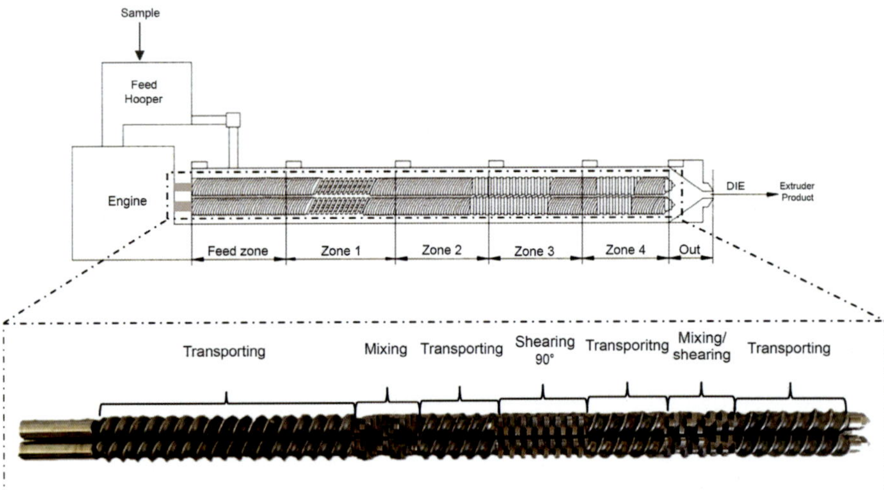


Figure 5.5. *An example of the extrusion process.*

*Source: Zambrano, Y., Contardo, I., Moreno, M. C., & Bouchon, P. (2022). Effect of extrusion temperature and feed moisture content on the microstructural properties of rice-flour pellets and their impact on the expanded product. Foods, 11(2), 198. https://doi.org/10.3390/foods11020198.*

## 5.4. FEED FORMULATION

For commercial purposes, feed formulation is the process of figuring out how much of each ingredient and additive to mix to make compound feeds that satisfy the known nutrient needs of the targeted species and fulfill production targets at the lowest possible cost. Two requirements are implied by the formulation of feeds:

- The necessary nutrients and the extent to which the targeted animals require them are known; and
- The following list of potential sources of these nutrients is appropriate for animal feeding. Based on these two guiding concepts, feed formulas can be improved even further by taking into account the functioning of ingredients, the impact of processing on raw material quality, the requirements of nutrients at different stages of life, the dynamics of production systems (such as ponds, cages, and recirculating aquaculture systems), and consumer expectations.

It may be argued that formulation techniques are less important to the effectiveness and quality of feed formulae than the breadth and precision of our understanding of these variables. The latter are merely information management tools. Feed formulation developed over time, moving from simple, but still used, manual calculation techniques based on the raw materials' crude nutrient content to more sophisticated computer algorithms that take into account the nutrient availability values of a wide range of components and additives. The feedstock data and nutrient list have grown significantly in the last few years. Rather than expressing ingredient amounts in terms of units required per metabolic body weight per day or other methods found in the scientific literature, commercial feed formulations commonly describe ingredient contents as a percentage (as fed or dry matter) for pragmatic reasons. In order to minimize the discrepancy between the expected and observed feed composition, it is also imperative to work on a consistent foundation (as fed or dry matter) when expressing the composition of ingredients. Feed formulation is driven by production goals. During the 1800s, common production objectives were weight gain, milk and manure output, or the repair of muscle tissue in terrestrial livestock animals. To accomplish these goals, the most nutrient-dense blends were created using data from constituent composition tables. The complexity of feed formulation techniques has evolved. Feed formulators can employ a variety of feed formulations that are easily found online and in textbooks, or they can apply more complex yet comprehensive methodologies like nonlinear programs, depending on their goals and level of technology access. Feed formulation techniques are becoming more and more dependent on the application of complex mathematical solutions in order to integrate the expanding body of knowledge in nutrition, diversify the objectives of production, and control risk and uncertainty regarding the availability of safe raw materials and the volatility of commodity prices.

## 5.4.1. Livestock Feed Formulation Methods

There are roughly six traditional approaches to creating animal feed.

### *5.4.1.1. Pearson Square Method*

The "Pearson Square" is a method that can be used to determine how much of two feeds an animal needs in terms of protein or energy. One of the feeds utilized needs to be higher in ME or CP than the desired level, and the other feed needs to be lower than the desired level, for this tool to function. The desired crude protein should be placed in the center of a square that has been drawn. Put the two grains in the square's left-hand corners, together with the amounts of crude protein they contain. Determine the difference between the values in the square's top and bottom left corners and the middle value. Then, arrange the result diagonally opposite. The percentage of each grain needed to create a grain mix with a CP level is represented by these figures. Determine the proportion of every grain. To get the total, add the two-part values. Then, figure out the proportion of each feed.

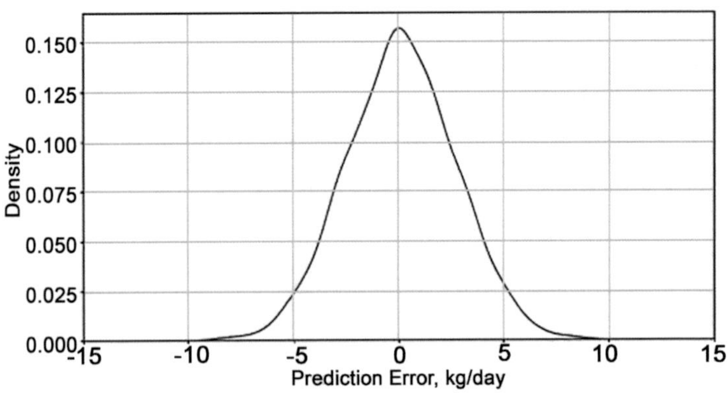


Source: Davison, C., Michie, C., Tachtatzis, C., Andonovic, I., Bowen, J., & Duthie, C.-A. (2023). Feed conversion ratio (FCR) and performance group estimation based on predicted feed intake for the optimization of beef production. Sensors, 23(10), 4621. https://doi.org/10.3390/s23104621.

This is a reasonably easy-to-follow and straightforward procedure. Among its benefits are its ease of use and its ability to balance protein requirements, to name just two. Its drawbacks include the fact that it can only be used for two (2) requirements at once and that other dietary requirements, particularly those for vitamins and minerals, are given less thought.

### *5.4.1.2. Simultaneous Equation Method*

This uses a straightforward algebraic equation as an alternative to the square approach. Here, two feed ingredients are combined to provide a specific nutrient demand. Both novices and seasoned feed millers will find the system easy to use. In teaching classes, it is used to expose pupils to feed formulation. The following are some benefits of the simultaneous equation over the Pearson method: First, a farmer can strike a balance between calories and protein. Second, it helps balance more complex rations by taking into account more than two feed elements at once. Ultimately, the system of equation grows as the necessity does. Its first drawback is that it only requires two feed ingredients and meets one nutrient requirement. An additional constraint pertains to the nutrient level computation, which necessitates an intermediate value between the concentrations of the two feed items being utilized.

### *5.4.1.3. Two-by-Two Matrix Method*

Using two distinct feed ingredients, this strategy addresses the needs for two nutrients. In order to find the answer to the problem, a set of equations is solved using a 2 by 2 matrix.

### *5.4.1.4. Trial and Error Method*

This is the most often used technique for creating rations for poultry and pigs. This kind of feed mixture is utilized worldwide in many underdeveloped countries, including Nigeria. As the name suggests, the formulation is adjusted until the animal's nutritional needs are satisfied. This technique enables the creation of a ration that satisfies the animal's complete nutrient needs. If care is not taken to thoroughly analyze or calculate the levels of calcium and phosphorus in the ration in question, various types of mineral deficiency, such as osteomalacia, rickets, and shellessness or soft-shell development, may not be adequately addressed in the production of chicken feed.

### *5.4.1.5. Imami Method*

This is a high-accuracy, instructive method for farmers without access to a computer to define and balance basic rations using a standard calculator.

### *5.4.1.6. Linear Programming (LP) Method*

This is also known as computerized feed formulation at the lowest cost. This approach uses a set of equations and linear programming techniques to find the least expensive combination of elements. It just requires seven simple actions to use this least expensive method in feed formulation.

Chapter 6

# Animal Feed Safety

## Contents

## 6.1. OVERVIEW

Animal feeds are essential to the global food chain because they make it possible to produce animal products profitably all over the world. They can be made using basic farm mixers or large feed mills. These feeds might be referred to as "industrial," "formula," "blended," or "compound." Under a variety of farming circumstances, animals are raised or kept for food, fiber, and other goods using manufactured feeds. Blended and balanced feeds are necessary for the efficient, intensive production of meat, milk, eggs, and other goods. By delivering sufficient nutrition at every stage of growth and production, safe feed solutions help farms maintain or improve food quality and consistency, lower production costs, and improve animal health and welfare. They can also lessen the possibility of pollution from animal excrement by supplying only the essential quantities of highly bioavailable elements in food. To maintain environmental safety, they should be utilized in concert with effective waste management systems that have been carefully thought out. Over a quarter of a million skilled laborers, technicians, managers, and professionals are directly employed in the commercial manufacturing or sale of manufactured feed products, which occurs in more than 120 countries. There are currently about 8,000 factories that can produce more than 25,000 tons of manufactured feed annually, in addition to additional production facilities like premix and specialty plants that can produce smaller quantities of high-value goods. All together, these facilities produce over 620 million tons of feed products a year. Worldwide, the commercial feed manufacturing industry is anticipated to have a turnover and sales value of US $85 billion annually. Global transportation is used to supply the sector with feed raw materials, branded products, handling and processing equipment, and technical services. The global feed industry is still growing in size and value due to factors including urbanization, rising consumer spending power, and population growth, even though it is currently serving mature, slower-growing industries in many industrialized nations. When personal income increases in nations where there is an unfulfilled need for more or better food, the demand for feed increases even faster. To address the global need for food that is both safe and reasonably priced, feed manufacturers must:

- Utilize a variety of technical and processing technologies in the production of feed, ranging from completely automated manufacturing systems to procedures that depend on skilled and general labor;
- Utilize a variety of co-products, by-products, and other excess raw resources from industrial, food, and primary agricultural sources;
- They carry out their own feeding and animal husbandry trials and sponsor university research in animal nutrition and other areas.

- Take on a more proactive role in educating consumers and engaging in global regulatory bodies' discourse on a range of matters on the availability of reasonably priced and safe animal-derived goods.

Adequate measures should be implemented at those points in the production and distribution chain where they will be most valuable in lowering feed-borne risks to consumers, according to a modern risk-based approach to feed safety. This ought to be mirrored in the use of particular precautions that are grounded on risk analysis and research, as well as a stronger focus on contamination management and prevention throughout processing. One important component is the application of the Hazard Analysis and Critical Control Point (HACCP) concepts. Risk-based initiatives have demonstrated efficacy in attaining the necessary level of hazard control for safeguarding consumers. Rather than being prescriptive and detailed, they are based on the desired result.

## 6.2. HEALTH HAZARDS ASSOCIATED WITH ANIMAL FEED

Animal feed can present chemical, physical, or biological risks to food safety. Each hazard has a specific source and pathway for exposure to contaminants, which must be thoroughly understood as the basis for risk management. It is important to recognize the potential risks associated with water, as hazards can be transferred from source materials to finished goods through handling, storage, shipping, or by-product contamination. Hazards may also arise from unintentional or intentional human activities, such as fraud or bioterrorism. Preparation and prevention should be the foundation of risk management, rather than merely responding after an issue has been identified. Significant challenges that have contributed to its development are as follows:

- Bovine Spongiform Encephalopathy (BSE) and other prion diseases;
- Impact on food safety from antimicrobial use in animals;
- Newly recognized undesirable substances: melamine, dioxins, dibenzofurans, and dioxin-like polychlorinated biphenyls (PCBs);
- Presence of genetically modified organisms (GMOs), crops, and enzymes in feed;
- By-products of new technologies (e.g., biofuel production) used in feed production;
- Radionuclides;
- Development of aquaculture industries and the search for new/better aquaculture feeds;
- Feed (and food) as targets of bioterrorism;

- Emerging technologies, such as the use of nanotechnology products in feed;
- Selection of undesirable substances and microorganisms of concern.

## 6.2.1. Chemical Substances

### *6.2.1.1. Dioxins, Dibenzofurans, and Dioxin-Like PCBs (Dioxins)*

Dioxins are present in the environment almost everywhere, therefore feed ingredients could potentially contain dioxins from a variety of sources. Since the 1999 Belgian dioxin crisis, dioxins have grown in importance when it comes to feed safety. Since then, a great deal of dioxin contamination from unanticipated sources has been documented. This has demonstrated that dioxins can be added during processing (like lime in citrus pulp) or be inherent in a product (like clay minerals). When treated wood, low-quality coal, or tainted fuel oil are utilized to dry feed products, dioxins may be released into the atmosphere. It has also been documented that dioxins can contaminate fodder crops that are cultivated close to specific industrial activities, such as incinerators. Dioxins and dioxin-like PCBs are two related classes of harmful substances that share several congeners. The toxicity equivalency factor (TEF) expresses the unique toxicity of each congener. It has been suggested that animal-derived foods, which may originate from the presence of dioxins in animal feeds, are the primary source of human exposure to dioxins. Due to the high degree to which dioxins accumulate in fat, even very low quantities of dioxin in feed have the potential to become significant during an animal's lifetime and to leave undesired residues in foods that are consumed by humans, such as meat, milk, and eggs. To assess the rates at which dioxins are transferred to animal tissues, toxicokinetic models have been created (Van Eijkeren et al., 2006). Therefore, putting controls in place for dioxins in feed is a critical first step in lowering dioxin levels in the food chain. Specifically, screening procedures have shown that dioxins can enter feed through the presence of mineral sources such clays, zinc oxide, recovered copper sulfate, and food by-products like fish meal and oils. The creation or enhancement of low-cost, precise screening techniques is required. In order to account for all sources of dioxin entering the feed chain, feed and food exposure studies are required.

### *6.2.1.2. Mycotoxins: Aflatoxin B1*

Numerous investigations on mycotoxins have been carried out in the past 10 years. At present, the effects of the most common mycotoxins (aflatoxin

B1, ochratoxin A, zearalenone, fumonisin B1, deoxinivalenol, T-2, and HT-2) on animal health are being studied. However, since food of animal origin represents a substantial route of exposure for humans, attention should be focused on mycotoxins that are known to move from feed to food of animal origin when examining the role that mycotoxins play in food safety. Aflatoxin B1 to liver, aflatoxin B1 to milk as aflatoxin M1, aflatoxin B1 to eggs as aflatoxicol, ochratoxin A to meat, deoxynivalenol to meat as DOM1, and zearalenone to meat as zearalenol are among the transfers from feed to food that the scientific community is aware of. However, the assessment of the rate and route of exposure in humans is limited to aflatoxin B1 for animals that produce milk. It is important for farmers to remember that animals given feed tainted with aflatoxin do not exhibit indications of aflatoxin toxicity. Cereals (especially maize), cottonseed, peanuts, copra, palm kernels, and rice bran are the feeds most prone to aflatoxin; however, care should be taken with all feed items grown in tropical and sub-tropical areas, especially if they are not dried or processed right away after harvest. Because aflatoxin contamination is not uniform, using the right sample technique is crucial. Feeds containing a high level of aflatoxin contamination shouldn't be given to animals that produce food or milk for human consumption, such as dairy cows. There are certain indications that mycotoxins may accumulate in soluble dry distillers' grains (DDGS) when grains are processed for the production of ethanol. They focus on cereal bran as well.

### *6.2.1.3. Heavy Metals*

Cadmium is a common contaminant found in many feedstuffs and feed components, especially minerals and forages that are farmed close to mining and smelting facilities. Heavy metals like arsenic and mercury are common in the environment and can be found in a variety of feeds, especially those that come from marine sources. Lead is another common pollutant. The most important minerals, their suppliers, and how they bioaccumulate in animal tissues are outlined below:

- **Arsenic (Inorganic):** Sea plants, fish products and supplemental minerals.
- **Cadmium:** Mineral supplements (such as phosphate, zinc sources). Forage/grains (depending on geographical area). Manure, sewage, sludge or phosphate fertilizers can enrich soil.
- **Lead:** Contaminated soil, lead paints, water from plumbing systems that contain lead, batteries. Mineral supplements (copper sulfate, zinc sulfate, zinc oxide), lead is also a natural contaminant of calcium carbonate (limestone) in some regions.

- **Mercury/Methyl Mercury:** Anthropogenic contamination, fish meal.

### *6.2.1.4. Veterinary Drugs*

Veterinarian medications should be used in accordance with good practices for the use of veterinary medications (GPVD) since they may pose a risk to food safety (Gupta, 2018). When animal products (both terrestrial and aquatic) are utilized in feed, there may be veterinary medication residues present; however, this is not a very major route of exposure. Due to the carryover of veterinary medications in feed during feed preparation, food items may contain veterinary drug residues. Therefore, while creating feed for animals that generate food, it is crucial to adhere to the Code's requirements about flushing, sequencing, and cleaning following the production of a medicated feed. The unlawful use of medicines in animal feed must also be considered, as this might leave dangerous residues in meat, milk, or eggs (such as chloramphenicol/nitrofurans in prawns and chloramphenicol in milk powder). Antibiotics that are employed in the fermentation process to manage microbial contamination when grains are processed to produce ethanol may concentrate in DDGS, according to some data.

### *6.2.1.5. Organochlorine Pesticides*

Organochlorine pesticides can be consumed through food due to their persistent usage in some nations and prolonged existence in the environment. This is because they can accumulate in the fat tissues of animals that have been fed contaminated feed. Typically, these animals won't show any particular clinical signs of the contamination. Meat and other animal products may contain these compounds, which break down very slowly and are incredibly tenacious. Human food safety concerns may arise from contaminated animal products.

### *6.2.1.6. Microbiological Hazards*

Contaminated pastureland, forages, and animal and vegetable protein meals supplied directly to animals are the main sources of microbiological dangers in feed.

### *6.2.1.7. Brucella*

In certain nations where Brucella infection is prevalent, ruminant carriers may give birth or abort in fields that are utilized for grazing or for harvesting grass for animal feed. It is commonly recognized that significant concentrations of Brucella microorganisms are present in the placentas of sick animals. Milking animals may expel microorganisms in their milk if they are fed polluted fodder. Food safety is in danger if this milk is not pasteurized before being consumed by people.

#### *6.2.1.8. Salmonella*

Salmonella continues to be a global health risk for people. It is evident that animal infections directly affect human transmission when animal products are consumed. Feed contamination may be a significant source of Salmonella exposure.

#### *6.2.1.9. Endoparasites*

Animal endoparasites that can infect animal feed include *Toxoplasma gondii*, *Cisticercus*, *Echinococcus*, and *Trichinella*. These endoparasites pose a risk to human health. These viruses have the ability to colonize and infect farm animals. If contaminated or infected products are consumed, there may be a health risk to humans.

#### *6.2.1.10. Toxic Plants*

Grasslands all across the world are home to a variety of poisonous plants. Their harmful consequences are widely known, as is the possibility that milk and meat contain some harmful substances (Riet-Correa and Medeiros, 2009). On the other hand, nothing is known regarding the average daily intakes (ADI), maximum residue limits (MRL), metabolic rates, and residues for these various toxicants. It is possible to manage this risk pathway by adhering to GAPs.

## 6.3. GENERAL PRINCIPLES AND REQUIREMENTS

For livestock production to be successful, feed and feed ingredient utilization must be appropriate, safe, and high-quality. To lessen and stop risks to food safety from entering the food chain, safe feed is crucial. Food safety risks that can result in issues with public health should be avoided or reduced when they are present in feed. Important tools for reducing risks in the feed manufacturing process include Hazard Analysis and Critical Control Points (HACCP), Good Manufacturing Practices (GMP), and Good Agricultural Practices (GAP) when applicable. The importance of feed safety has been emphasized by the advent of the food chain concept, which acknowledges that accountability for the provision of safe, wholesome, and nutritious food is shared along the entire food chain. Therefore, the entire process from initial production to ultimate consumption is included in the food chain. Farmers, fishers, operators of slaughterhouses, manufacturers of feed ingredients, producers, and processors of feed, food processors, transport operators, distributors (wholesale and retail), and consumers are among the stakeholders. Governments tasked with safeguarding the public's health are also involved. To lower the danger to human health, all parties involved in the production of feed and animals should make sure that high-quality, safe feed and feed ingredients are produced and used in food-producing animals. To establish the relationship between any identified or possible hazards and the level of risk, cooperation is required from all parties involved in the food production chain, including those who are able to give veterinary clinical and epidemiological information. Safe feeding procedures and the development and upkeep of suitable risk management strategies depend on this kind of information. A set of guidelines known as the Codex Alimentarius Code of Practice on Good Animal Feeding is designed to make sure that feed and feed ingredients are sourced, manufactured, processed, transported, stored, distributed, and used in a way that does not pose a risk to human health.

### 6.3.1. Feed Ingredients

For the production of safe and high-quality feed—which is necessary for the production of safe and high-quality animal food products like meat, milk, eggs, and so on—quality and safety of feed ingredients are paramount. Before using feed ingredients for animal feeding, their safety should be evaluated. The Codex Principles for Risk Analysis should serve as the foundation for the evaluation of feed and feed additives. Relevant Codex texts, such as the Principles and Guidelines for the Conduct of Microbiological Risk Assessment, the Risk Analysis Principles applied by the Codex Committee on Pesticide Residues, the Risk Analysis Principles applied by the Codex Committee on Residues of Veterinary Drugs in Foods, the Risk Analysis Principles applied by the Codex Committee on Food Additives, and the Codex Committee on Contaminants in Foods, should be taken into consideration when developing the risk assessment of microbiological and chemical hazards in feed and feed ingredients. To prevent the entry of food hazards into the food chain, information should be supplied to guarantee that feed and feed ingredients are utilized and stored properly. Users of feed ingredients should confirm that the ingredients they bought are free of contaminants that processing would not normally be able to eliminate. Product details enable the following:

- The reduction of losses by putting in place effective recall protocols.
- Improved quality and process control as a result of raw material information being readily available.

- Measurements should not be repeated needlessly in two or more phases.
- Potential to link processing and raw material properties with product data.
- Improved planning to maximize each product type's use of raw materials.
- Avoiding the unprofitable blending of inferior and superior raw materials.
- Accessibility of information during audits of quality management.

The formulation of ingredient specifications is crucial to the execution of the program for quality and safety assurance. The agreements with suppliers, the feed composition, the hazard analysis, and the controls that follow are all based on specifications. Buyers' ought to assess vendors according to their capacity to deliver goods in compliance with predetermined requirements. Official ingredient definitions may be used in purchasing specifications, which should be developed to precisely define the good or service that needs to be ordered. Purchase contracts, supplier certification, supplier visits, ingredient supply monitoring, and a mix of these methods can all be used to assess suppliers. The feed business uses a wide range of technologies across the globe to ensure the quality and safety of the different feed ingredients. Some nations have lists of ingredients that are prohibited from use, lists of ingredients that can be used with restrictions, lists of ingredients and their amounts that are excluded, and lists of ingredients that are allowed to be used with restrictions or for specific purposes. It is important to sample feed ingredients frequently to make sure that quality and safety requirements are being fulfilled. Health issues related to animal feeding will be reduced with regular housekeeping and testing for any suspected pollutants. If a feed ingredient is suspected of being contaminated, it should not be used to make animal feed until it has been thoroughly tested and determined to be suitable for the species and class of animal for which it is intended.

#### *6.3.1.1. Labeling*

To keep health risks out of the food chain, product labels should give consumers all the information they need to handle, store, and utilize feed and feed additives safely. It's critical that users have sufficient training to comprehend labelling information completely and apply it effectively. Users can address the dietary requirements of animals based on their physiological and productive demands by providing them with information about the contents and purpose of the feed. Because the risk to human health may vary depending on which species or animal categories the feed is meant for, it is required to label the species and animal categories for which the feed is intended (e.g., mammalian proteins when fed to ruminants). Inadequate product information, along with a lack of understanding about food hygiene and general feed, might result in improper handling of items later on in the food chain. Even in situations when sufficient hygiene control procedures have been implemented earlier in the feed chain, such deception may lead to feed contamination or products that are no longer fit for human consumption. For the purpose of record keeping, traceability/product tracking, and product recalls if needed, all labelling information regarding the source of feed and feed additives (such as manufacturers, lot identification, manufacturing date, etc.) is crucial. Effective stock rotation may also be aided by this knowledge. Accurate labeling ensures that working inventory, packing, and other records receive accurate information. Specific details on the active medication ingredients, the kind and class of animals for whom the feed is meant, the purpose or indications of usage, and warning and caution statements must all be included on the label of medicated feed. The withdrawal periods and other phrases pertaining to the preservation of human health are examples of warning statements. The abuse of medicated feed and concerns about drug stability and animal safety are the subjects of cautionary remarks.

### 6.3.1.2. Traceability and Record-Keeping

When and when it is appropriate, traceability/product tracing can be used in the feed and food chain to protect human health from food-borne illnesses and dishonest marketing tactics. It can also help to facilitate commerce by providing correct product descriptions. At any point in the feed and food chain, traceability, and product tracing make it feasible to determine the origin and destination of the product (feed or food). This would make it possible to compile a set of historical data that would track a product all the way through manufacturing. Appropriate record-keeping practices that trace a specific products or ingredient's journey from suppliers to the company, via all intermediate stages that combine and process ingredients to create new products, and to customers via the supply chain, facilitate traceability and product tracing. The ability to identify a particular product at every stage of the feed and food chain is the foundation of traceability and product tracing. New identities are continuously formed as products are bulked up for delivery, recipes are mixed, and/or huge batches are divided to multiple locations throughout the feed and food chain. For a batch to be considered traceable, it must be able to be identified and provide ties to the product's past. It is possible to carry additional information, such as details on ingredient quality or origin or processing efficiencies that need to be calculated for industrial systems. The type and quantity of data can be expanded as the system dictates, and it can be stored for a portion of the food chain or the entirety of it. When appropriate, traceability, and product tracing can be utilized to support product recalls or withdrawals as well as non-conformity investigations.

### *6.3.1.3. Conditions for Emergency Situations*

To reduce any potential negative impacts on public health, prompt notification of the kind and severity of the safety issue to all pertinent parties is crucial in the event of a feed emergency. Experience has proven that to guarantee food safety, data regarding feed and food safety situations need to be combined into a unified system. A system like this ought to include standards for recognizing emergency scenarios. When a hazard (such as contamination) is discovered, the responsible authorities should investigate its source and, if feasible, take the necessary action to minimize or eradicate it. Traceability and product tracing are crucial tools for quickly determining the source of dangers in emergency situations.

### *6.3.1.4. Inspection and Control Procedures*

Feed operators and appropriate authorities share responsibility for producing safe and high-quality feed and feed additives. Feed operators and responsible authorities must adopt efficient control methods and feed inspection programs to ensure feed safety and quality. The impression of control measures' success among feed users influences their trust in the quality and safety of their supply, and eventually, that of consumers regarding the food's quality and safety. Self-control programs help feed operators meet additional requirements (such as manufacturer or purchaser-defined specifications) and appropriate regulatory norms. Programs for self-control should cover intermediates, final feed, and incoming feed. Physical inspections, sample techniques, chemical and microbiological analyses, actions taken in the event of noncompliance, staff responsibilities related to production and feed safety management, and so on, are some examples of self-control systems. To ensure adherence to legal obligations, regulated feed inspections must be conducted by competent authorities. To ascertain whether a company is essentially in conformity with the regulations, surveillance checks are carried out. Compliance inspections are carried out to assess whether a company is adhering to the regulations' requirements and to record inspectional findings that may be used to justify enforcement action. Targeting businesses, establishments, goods, and procedures that pose the most threats to the health of people or animals through the use of a scientific and risk-based approach will enhance the capacity to prioritize and distribute inspection resources. The risk and the dependability of the controls currently implemented by those handling the items, such as producers, manufacturers, importers, exporters, and distributors, should be considered when designing the frequency and strength of inspection system controls. The product's danger to human health and safety, its place of origin, its history of compliance with regulations, and other pertinent data should all be taken into consideration when determining the type and frequency of inspection, sampling, and testing. The risk that the product poses to human health, the possibility that the requirements won't be followed, and the track record of compliance of the producers, processors, manufacturers, exporters, importers, and distributors should all be taken into consideration while designing controls. A crucial component of any program for quality assurance and control is laboratory testing. To ensure that a feed or ingredient sample satisfies quality standards, certain components are measured. Tests evaluate a product's quality against a predefined standard by measuring its biological, chemical, and physical attributes.

### *6.3.1.4.1. Health Hazards Associated with Animal Feed*

Animal feed can pose chemical, physical, or biological risks to food safety. Each hazard has a specific source and pathway of exposure to contaminants. Risks can be transferred from source materials to finished products through handling, storage, shipping, or by-product contamination. Hazards may also arise from unintentional or intentional human activities, such as fraud or bioterrorism. Feed-borne diseases may result from contaminants such as mycotoxins, excessive veterinary medicine residues, industrial and agricultural pollutants (e.g., dioxins), and infections, such as the agent causing bovine spongiform encephalopathy (BSE).

### *6.3.1.4.2. Feed Additives and Veterinary Drugs Used in Medicated Feed*

Any combination of veterinary medication(s) and feed(s) that is ready for commercialization and meant to be fed to animals raw is known as a medicated feed. This is because the mixture has additional medical qualities in addition to its curative and preventive qualities. Veterinarian medications produced ahead of time to later produce medicated meals are known as premix for medicated feeds or medicated premixtures. Veterinary medications should be used in accordance with best practices to avoid any potential risks to food safety. Antimicrobial medications are effective instruments for treating infectious infections in both people and animals. All nations must set up the proper mechanisms to guarantee that veterinary antimicrobial medications are produced, marketed, supplied, prescribed, and used responsibly. These mechanisms also need to be sufficiently audited. Feed additives are added to feed for a variety of reasons, such as enhancing digestibility, improving physical and organoleptic qualities, enhancing palatability, extending shelf life, preventing deterioration, and modifying the color of egg yolks and salmon trout, among other characteristics. Several materials, including vitamins, enzymes, and microbes, can be categorized as feed additives based on how they are administered and for what purpose. Feed additives must be developed and used in compliance with applicable laws and production guidelines, as well as safety evaluations.

### *6.3.1.4.3. Feed and Feed Ingredients*

Animals that produce food may be contaminated by feed, feed additives, or forage. Foods of animal origin may contain chemicals and biological materials that are added to feed, either knowingly or unknowingly, at various points in the feed manufacturing process and given to the animals. Feed and feed ingredients must come from reputable sources, be stored under stable conditions, and be handled carefully to avoid harmful effects from contamination or degradation. Feed and feed ingredients must meet all applicable safety and quality criteria and be in good condition when they are manufactured or received. For the concentration of unwanted compounds in food intended for human consumption to continuously remain below the level of concern, all levels of these substances must be sufficiently low in feed and feed ingredients. Food-borne dangers that come from feed should be sufficiently regulated. To guarantee consumer safety, quality assurance is necessary throughout the entire production process. It is the responsibility of manufacturers to supply sufficient information that will allow feed quality and safety to be preserved after delivery. Controls should be implemented at every stage of the production process to find any possible health risks related to the product. These measures ought to guard against contamination of both finished and incoming feed. Feed and feed additives that include unacceptable amounts of unwanted substances should be marked clearly as unfit for animal feed and should not be sold or utilized. In order to reduce the chance of contamination, intake points, processing machinery, conveying systems, and storage facilities should all be built and run.

### *6.3.1.4.4. Undesirable Substances*

Pathogens, mycotoxins, pesticides, industrial and agricultural chemicals, heavy metals, and radionuclides are a few examples of undesirable substances. It is important to locate, regulate, and reduce any undesirable compounds that might be present in feed and feed ingredients. Unwanted substances must be cut down to manageable amounts that don't have negative side effects. The techniques for identifying undesired substance residues are getting more advanced, allowing for the detection of even minute amounts of residues that are safe for both human and animal health.

### 6.3.2. Good Production Practices

There is a growing need for everyone involved in the production and use of feed to pay more attention to risk management because of the consequences for human health. It is imperative that feed manufacturing and feed production in general be treated as significant links in the food production chain, given the clear correlation between feed safety and the safety of food products derived from farmed animals. "All conditions and measures necessary to ensure the safety and suitability of feed at all stages of the feed production chain" is how the Code defines feed safety. Several national feed associations have created safety initiatives based on the idea that feed is already widely regarded as a component of the food chain in several countries. They all highlight how crucial it is to maintain feed products contaminant-free and safe. Risks to feed and food safety can arise at any point in the feed processing chain, hence it is crucial to have enough control over the entire feed and food chain. A significant number of international governments, trade associations, and private sector organizations have released standards, guidelines, and legislation that highlight the obligation of all parties involved in the food chain to guarantee feed safety. The major objective is to guarantee that the risk of contaminated feed for animals that produce food is maintained as low as possible. Two important tools in achieving this are the use of HACCP principles and the maintenance of traceability. Important preparatory programs for putting HACCP concepts into practice are GMPs and GAPs. The HACCP plan in conjunction with preparatory programs ensures effective hazard control.

#### *6.3.2.1. Good Manufacturing Practices*

GMPs, or good manufacturing practices, are methods and protocols that guarantee the appropriateness and safety of food and feed and ought to be used all the way up the feed chain.

#### *6.3.2.2. Buildings and Facilities*

All facilities and buildings should be designed and built with the protection of feed products against contamination in mind at all times. Enough room should be available for all activities as well as the secure storage of tools and supplies. For the purposes of cleaning and maintenance, easy access should be feasible. The location, layout, and structure of a building should discourage pests and minimize their ability to enter.

### *6.3.2.3. Location of Feed Establishment*

Where feed establishments are located should take into account potential sources of contamination as well as the efficacy of any reasonable precautions that may be implemented to protect feed. Businesses should be situated away from unfavorable concentrations of dust, smoke, and other pollutants. Generally speaking, establishments should be situated distant from:

- Industrial operations and places with high environmental pollution are major risks for polluting feed.
- Places that flood easily (unless adequate protections are installed).
- Locations that are vulnerable to pest infestations or the presence of both domestic and wild animals.
- Regions where it is not possible to properly remove wastes, either liquid or solid.

### *6.3.2.4. Design and Layout*

Establishments should be able to implement excellent hygiene measures, such as cross-contamination prevention, through their interior structure and design. Activities where cross-contamination could occur should be sufficiently isolated by physical barriers or other practical measures. Structures and amenities ought to be planned to facilitate effortless cleaning access, encompassing entry to the inside of pertinent machinery. Enough room should be available to carry out all process activities and product inspections in an efficient manner. The outside of the structure should be planned, built, and maintained to keep pollutants and pests out. No exposed apertures should exist, air intakes must be positioned correctly, and leaks must be prevented by maintaining the base, walls, and roof. Only the outside spaces should be used for gardens and other vegetation. Parking lots, outdoor spaces, and all access points to the manufacturing facility should be planned to prevent contamination of the production area—for instance, from cars dragging snow or mud into the area. When necessary, separate, well-designed storage spaces should be provided for hazardous, explosive, or combustible items. These spaces should be kept apart from areas used for manufacturing, storage, and packing. It is important to protect the safety of incoming raw materials and departing final feeds while designing and building intake and loading facilities. It is important to have controls in place to prevent water or pest contamination.

### *6.3.2.5. Internal Structure and Fittings*

The establishment's structures ought to be made of sturdy materials. They ought to be simple to clean, maintain, and, when necessary, disinfect. When necessary, the following requirements in particular should be met to guarantee the feed's compatibility and safety:

- The surfaces of floors, walls, and partitions should be composed of non-toxic, impermeable materials when used as intended.
- Smooth surfaces on walls and dividers make cleaning easier and more possible.
- The design of floors should provide sufficient drainage and cleaning capabilities as required because of the way they operate.
- It is best to build and finish ceilings and overhead fixtures in a way that minimizes particle shedding and the accumulation of dirt and humidity.
- Windows should have detachable, washable insect-proof screens, be easy to clean, and be designed to reduce dirt accumulation.
- Doors should be easy to clean and have smooth, non-absorbent surfaces.
- Work surfaces that come into direct touch with feed ingredients, including weighing tables, should be robust, easy to clean, and maintain.

### 6.3.2.6. Water Supply

Water that comes into contact with feed items ought to be potable. Potable water should be available in sufficient quantities and be equipped with the right infrastructure for distribution, storage, and temperature control. Potable water should meet the standards outlined in the most recent WHO drinking water quality guidelines. A separate system should be used for non-potable water used for refrigeration, steam production, fire control, and other similar uses. It is important to identify non-potable water systems and to avoid connecting to or permitting reflux into potable water systems. Every hose, tap, and other such potential source of contamination needs to be made to stop siphoning or backflow. Chemicals used in water treatment should be compatible with food when employed. To guarantee that the right dosage is administered, chemical treatment needs to be watched over and managed. Recirculated water needs to be kept, inspected, and treated appropriately for the use for which it was designed. There should be a distinct and easily identifiable distribution system for recirculated water.

### 6.3.2.7. Cleaning Facilities

Suitably designated and ample facilities should be available for cleaning feed utensils, equipment, and feed product-transporting vehicles. When necessary, these facilities ought to have a sufficient supply of both hot and cold water. Buildings should ideally be made of readily cleaned corrosion-resistant materials and have potable water available at a temperature suitable for the cleaning agents being employed. Every cleaning agent needs to be safe for food use. To avoid contamination, areas used for processing, packaging, and feed storage should be sufficiently isolated from equipment cleaning facilities. To guarantee that an acceptable level of personal hygiene can be maintained, personnel hygiene facilities should be made available. Facilities that should be available include:

- Sufficient methods for properly cleaning and drying hands, such as sinks with hot and cold water or water with a temperature that is appropriately adjusted.
- A steady flow of drinkable water.
- A sufficient quantity of sanitary restrooms with nearby handwashing stations that are stocked with soap, paper towels, or other acceptable hand-drying materials.
- Sufficient facilities for staff to change in.

Facilities ought to be placed and designed appropriately. Facilities for hand washing and/or hand disinfection should be available in product handling areas whenever the nature of the business demands it.

### *6.3.2.8. Air Quality, Temperature, and Ventilation*

Adequate means of natural or mechanical ventilation should be provided to:

- Reduce the amount of aerosol and condensation droplet airborne contamination of feed, particularly in open production systems.
- In areas where ambient temperature might have a negative impact on feed safety, regulate it. Systems for heating, cooling, and air conditioning should, if required, be built and installed to prevent product, equipment, or utensil contamination from air intake or exhaust vents.
- Make sure there is enough airflow to keep moisture and oil from building up on the walls and ceilings.
- Maintain humidity levels while making sure the food is suitable and safe.

Systems for ventilation should be built and configured to guarantee that only clean air is drawn into intakes. Air should ideally move from clean to contaminated places according to design. Systems for mechanical ventilation need to be cleaned and maintained properly.

### *6.3.2.9. Lighting*

Sufficient lighting sources are necessary to maintain hygienic conditions in the production and storage facilities, hand-washing areas, toilets, and spaces where equipment and utensils are cleaned. When artificial lighting is necessary, it ought to be made with the intention of reflecting genuine color. Conditions for adequate lighting are especially crucial in locations where instruments are monitored or feed is visually reviewed.

### *6.3.2.10. Equipment*

Nontoxic materials should be used to make equipment and containers, and they should be able to be disassembled for regular maintenance, cleaning, and inspections. In order to enable cleaning and maintenance and to avoid insect infestation, equipment should be positioned away from the walls. Appropriate metering devices should be installed on equipment intended to attain and regulate particular process parameters, such as temperature, humidity, and air flow, and their accuracy should be routinely verified. The purpose of these specifications is to guarantee that:

- Harmful or undesirable micro-organisms or their toxins are eliminated or reduced to safe levels, or their survival and growth are effectively controlled;
- Where appropriate, critical limits established in HACCP based plans can be monitored;
- Temperatures and other conditions necessary to feed safety and suitability are achieved and maintained.

Waste, byproduct, and hazardous or inedible material containers need to be clearly marked and built appropriately. It is imperative to identify and secure containers containing hazardous materials to avert product and environmental contamination. Feed items should never be kept in containers intended for hazardous goods or garbage. Spoons and knives, for example, which are used to weigh pharmaceuticals and additives, and open bags should be tethered or otherwise secured and should not be left on the ground or on top of pallets or raw material bags. The variety of weights and volumes needed to create homogenous mixtures must be accommodated by mixers. Scales and other metering devices are examples of weighing equipment that has to be suitable for the weights and volumes that will be utilized. The dose and weighing apparatuses' accuracy should match the objects to be weighed. Controls should be in place whenever bulk bins are used to guarantee that only the appropriate raw materials are placed into any bin. Screens, filters, separators, and sieves should all be inspected frequently to make sure they are functioning properly and to look for potential damage. When utensils, containers, and other equipment come into touch with feed, they should be made to be able to be properly cleaned and maintained when needed to prevent feed contamination. Utensils, containers, and equipment should be manufactured of materials that won't be harmful when used as intended. Equipment should be made to be able to be cleaned, maintained, monitored, and made easier to inspect for pests. Paints, coatings, lubricants, chemicals, and other materials applied to surfaces or equipment that might come into contact with feed should be chosen so as to minimize the risk of unacceptably high feed contamination. Feed mixing, cooking, storing, and transportation equipment need to be built to meet and maintain the necessary operating conditions. Essential temperature, humidity, pressure, and mixing conditions should be able to be monitored and regulated with such equipment. Any controls put in place ought to guarantee that the following:

- Critical limits specified in HACCP-based strategies can be tracked when necessary.
- It is possible to easily establish and maintain the temperature, humidity, and other process parameters required for feed safety and appropriateness.

The manufacturers' recommended calibration techniques and frequencies should be followed for any equipment that monitors or controls feed safety. Equipment calibration should only be done by those with the necessary training.

### *6.3.2.11. Personal Hygiene*

If there is a chance that someone could contaminate feed products, they shouldn't be permitted to enter any processing areas if they are known or suspected of having a sickness or ailment that could be spread through the feed. Anyone who is impacted by this should notify management right away if they become ill or exhibit any symptoms, and they will either be given appropriate work or sent home. The following symptoms need to be mentioned to management:

- Jaundice;
- Diarrhea;
- Vomiting;
- Fever;
- Sore throat with fever;
- Visibly infected skin lesions (boils, cuts, etc.);
- Discharges from the ear, eye or nose.

Feed handlers are required to wear sanitary safety boots, proper protective clothes, and a head covering when necessary. They should also maintain personal hygiene. In addition to providing protection for staff when needed, clothing should be made to prevent personnel from contaminating feed products. Controls should be in place when gloves are worn to make sure they don't get into the feed products. Clear guidelines on eating, drinking, and smoking should be posted on the premises. Away from locations where feed products are handled, stored, or processed, there should be designated facilities available. Personal belongings shouldn't be brought into places where feed is handled, processed, or stored because they could fall out of pockets and endanger the feed's safety and suitability. When visiting the processing and handling facilities, contractors, and other visitors, including employees, should wear protective clothes and follow the other personal hygiene guidelines.

### *6.3.2.12. Cleaning*

The remaining dirt and residues should be cleaned up since they could be a source of contamination. Feed products must be compatible with cleaning materials and techniques. To minimize the risk of exposure to pests and pathogens during the processing, storage, and handling phases, it is imperative to implement adequate standards of cleanliness. Cleaning procedures that guarantee processing, storage, and handling facilities are kept sufficiently clean to guarantee feed safety should be recorded. Programs for cleaning and disinfection should be evaluated for appropriateness and efficacy. Cleaning inspections should be conducted by an authorized individual, and a log of each inspection should be maintained. Feed items should only be cleaned and disinfected/sanitized using food-compatible chemicals, and they should be used in compliance with safety data sheet specifications and manufacturer guidelines. When cleaning supplies and sanitizers come into contact with feed items, it's important to make sure control systems are always providing the right amount of dilution. Chemicals used for cleaning and disinfection/sanitization must, when needed, be stored separately in containers that are easily identifiable to reduce the possibility of contamination—either intentionally or accidentally. Upkeep A routine of scheduled maintenance should be applied to equipment to guarantee that it is kept in a functional and safe state. Any maintenance performed on equipment that is vital to producing safe feed, such as magnets, cookers, and crucial measuring tools, should be documented. On-site engineers and contractors should be managed to prevent building and maintenance projects from jeopardizing feed safety. When operations are resumed in areas where building or maintenance work has been done, a protocol should be in place to guarantee that the necessary cleaning and tidying has been finished.

### 6.3.2.13. Pest Control

In all locations used for processing, storage, and handling, proactive steps should be taken to restrict and manage pest activity. Whether an animal is domestic, wild, or feral, risk assessment techniques should be utilized to find possible issues with all animal classes (e.g., birds, insects, reptiles, and mammals). To demonstrate that pest threats are regularly and effectively addressed, records should be kept. Whenever possible, animals should be kept off feed manufacturing facility grounds as well as the areas around retail businesses and processing facilities. In situations where pest presence is inevitable, precautions should be taken to shield feed products from possible contamination. Access points should be proofed against pest invasion whenever there is a substantial risk of pest infestation. Doors must fit tightly and be pest-proof when closed, and they should be maintained closed whenever feasible. To stop pests from entering buildings and to get rid of possible breeding grounds, buildings should be maintained in good repair and condition. Whenever feasible, holes, drains, and other areas where pests could enter should be plugged. When sealing isn't an option, there should be precautions taken to lessen the likelihood of insect ingress, like installing wire mesh screens. Any action performed in response to a pest infestation should be compatible with feed products and should be done so quickly. The only people who should perform any necessary control treatments are those who are suitably qualified and trained. Lead or other harmful ammunition should not be used when shooting is done as part of a pest management program. Unless there's a compelling reason not to, all bait containers should be fixed in their proper positions. It is not advisable to place open bait containers or loose baits in locations where they could pose a risk to feed products. It is important to keep records of pest management operations and make sure that feed products are not contaminated with substances intended to eliminate or discourage pests. Records for pest control should contain:

- Details of any poisons used, including safety data sheets;
- Qualifications of personnel involved in pest control activities;
- Map(s) indicating the location of any bait stations and the baits with which they are baited;
- Records of any pests found;
- Details of corrective actions implemented.

### 6.3.2.14. Waste

Waste and materials that shouldn't be fed to animals must be recognized as such, stored separately, and removed. In areas used for handling, processing, and processing feed, waste should not be permitted to build up. To reduce the possibility of unintended or inadvertent use, waste should be gathered, stored, and labeled bins or containers. Trash must be disposed of legally and in compliance with any environmental laws that may be in force. Feed products should not be placed in waste disposal containers. Cover the containers you use to keep rubbish that attracts bugs. Additionally, these trash containers must be kept out of processing and storage facilities and taken down from the site as often as is reasonably possible. Waste management facilities need to be included in cleaning and disinfection programs and need to be kept properly clean. Drains: Every drain needs to be planned and kept up so that it doesn't pose a risk to any feed items. Feed components should not contain any wastewater or material collected from wastewater systems.

### *6.3.2.15. Storage*

To avoid cross-contamination, raw material and end product storage facilities should be kept apart. Pesticides, fertilizers, chemicals, and other possible contaminants ought to be absent from these facilities. Feed items need to be stored such that they are easily recognized and don't get mixed up with other products. Pre-mixtures and medications should be kept in a safe location with limited access for authorized staff only. To avoid unintentional use, any rejected products should be properly marked and stored in designated areas. Approved and specification-compliant finished feeds ought to be kept in appropriate containers or packing materials. Feed containing medication needs to be kept in a safe, separate location. It is important to build storage facilities with pest prevention in mind. Regular cleaning and clearing of storage areas is necessary. To stop the growth of mold, raw materials and completed goods should be stored in a cold, dry environment. When needed, humidity and temperature should be regulated. Enough stock control mechanisms should be in place to guarantee that neither completed feeds nor raw materials degrade before being used, shipped, or stored. Raw resources must be used whenever possible, and feed materials must be provided on a first-in, first-out basis.

### *6.3.2.16. Transport*

Throughout transportation, sufficient protection for both raw materials and completed feeds is required. To reduce and limit the risk of contamination, all modes of transportation—owned or leased, bulk or packed, by land, rail, or water—should be suitably cleansed. The type of loads being transported will determine the best cleaning technique. Generally speaking, the load compartment should be kept dry, and when possible, vacuuming or sweeping should be done. Use of a steam cleaner or pressure washer will be required where products that are sticky or damp are being moved. Before being utilized again for the transportation of feed products, vehicles used for the transportation of medicated feed and other high-risk items (including those later found to be infested with diseases or insects) should be thoroughly cleaned, sanitized, and dried. Contractual transportation needs to be carefully monitored, and rental agreements ought to include the upkeep of clean transportation. It is important to routinely verify that this requirement is being followed. Before the feed items are loaded, nothing from the prior loading should be left in the tank trucks, crates, or other containers. Dry and clean containers are necessary before filling. Verify that the loads that were previously transported in any transport are compatible with the load that is being fed next. It is important to verify the three prior loads that were transported and obtain assurances that no feed-related material was transported using a vehicle that might have contaminated the area over time. To guarantee hygienic transportation conditions and the absence of residual material buildup, all vehicles utilized for the transportation of feed items ought to undergo routine cleaning and sanitization procedures. Products need to be kept dry and shielded from contamination. Loads should be covered when transporting in closed vehicles is not an option. Additionally, the cover needs to be kept dry, clean, and sanitized.

#### *6.3.2.17. Training*

To guarantee that feed and food remain safe, proper training is necessary. Feed hygiene, production procedures, and feed product handling should all be covered in training for those working in the manufacturing and processing of feed. Every employee needs to understand their own roles and responsibilities in ensuring feed safety. Every training session needs to be recorded. Regular supervision and checks to make sure procedures are being followed correctly should be conducted, as well as periodic evaluations of the success of training and education programs. To assess any hazards and take appropriate action, managers, and supervisors should be well-versed in feed and food hygiene concepts and procedures. Programs for training should be revised and assessed on a regular basis.

### 6.3.3. Hazard Analysis and Critical Control Points (HACCP)

The goal of a formal hazard analysis should be to identify and control any hazards that could compromise the safety of feed products for human consumption. The "Recommended International Code of Practice General Principles of Food Hygiene CAC/RCP 1-1969, Rev. 4-2003" published by the Codex Alimentarius Commission defines the internationally recognized HACCP preliminary phases (items 1–5) and principles (items 6–12). Because HACCP is plant-specific, companies that produce comparable feeds can use different HACCP plans in terms of the hazards they identify and the controls they put in place.

### *6.3.3.1. Prerequisites*

The HACCP team should have fundamental operating processes in place that have been verified as effective by internal auditing systems before starting to design a HACCP plan. For the HACCP, these actions are known as "prerequisites"—that is, "needed as a prior condition." Prerequisite programs include the following; examples of them include the previously mentioned Good Manufacturing Practices:

- Smoking, eating, and drinking policy;
- Cleaning schedules and hygiene audits;
- Pest control program;
- Supplier approval procedures;
- Plant operating procedures and instructions;
- Equipment maintenance;
- Job descriptions and responsibilities;
- Staff training.

The HACCP team can concentrate on dangers that cannot be managed by other means when the effectiveness of precursor programs to control possible hazards is validated. Prerequisites and the HACCP plan itself must be reviewed in subsequent evaluations to guarantee that process modifications and recently unknown risks are managed. For the HACCP plan to be applied and implemented successfully, all necessary programs—including training—need to be well-established, completely functional, and validated.

### *6.3.3.2. HACCP Team*

Establishing a HACCP Team is the first preparatory step in creating a HACCP plan. Employees from all pertinent departments and operations within the organization should be on the HACCP team, along with at least one person who has received HACCP training that has proven to be successful. The HACCP Plan should list each member of the HACCP team. As long as the team's function is maintained, it is allowed for a single employee to play more than one job within the HACCP team or to use resources from outside the organization. The following disciplines will be represented in a traditional HACCP team, while they may not always be represented by separate individuals:

- **Team Leader:** This person, who might be among those listed below, should preferably have received training in HACCP concepts and have hands-on application experience.
- **Quality Assurance/Quality Control/Technical:** Someone who

is knowledgeable with the items being considered, their historical risks, and the important concerns surrounding them will be needed for this.

- **Production:** Someone with extensive knowledge of the specific steps involved in the production process and close involvement in the process is needed for this.
- **Engineering:** Someone knowledgeable in processing plant mechanics, material accumulation inside machinery, potential heat or moisture applications, and machinery access procedures will be needed for this.
- **Additional, Part-Time Expertise:** Specialists with technical or specialized knowledge in areas like distribution, microbiology, buying, operational tasks, and requirements for certain species can be needed for this.

Team members must be somewhat involved in day-to-day operations and knowledgeable about what really occurs in the company.

### *6.3.3.3. Product Description and Intended Use*

The HACCP team first provides a written specification for the feed product, which includes an overview of the product's contents, intended purpose, and general description. Distribution of the product may be done through retail, wholesale, internal use, or additional processing. The typical expected use of the product is also described by the HACCP team. During the hazard analysis stage of developing a HACCP plan, this data is employed. Any relevant changes should be reflected in the published product specifications. The written specification created for the HACCP plan should also contain the following:

- Nutritional and analytical characteristics;
- Hazards or limitations for intended use, where these apply;
- Details of any medications included; and associated withdrawal period.

### *6.3.3.4. Definition of Process Steps*

Every step in the HACCP team's activities should be identified and documented, starting with the procurement of raw materials and supplier approval, and ending with the transfer of any produced feeds to a buyer (process analysis). A flow diagram illustrating each stage of the process operation should be used to support the process analysis.

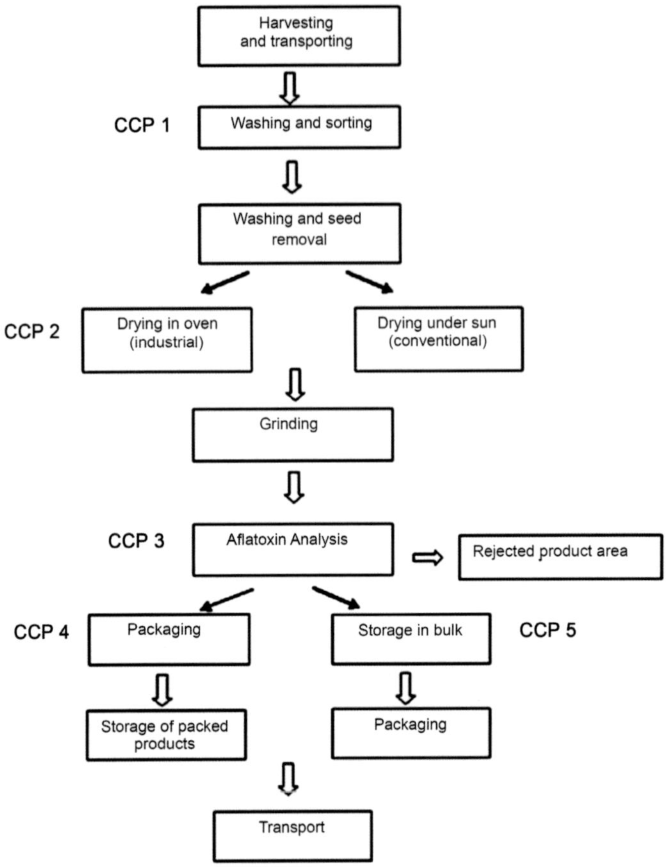


**Figure 6.1:** *Sample HACCP flow chart with CCP (critical control point)*

*Source: Ozturkoglu-Budak, (2017).*

### *6.3.3.5. Hazard Analysis /Identification (Codex Principle 1)*

The HACCP team should enumerate every potential risk that might possibly be anticipated to pose a threat at each stage of the procedure. All risks ought to be enumerated at this point, and those that might be excluded from the investigation in the future should be noted as prerequisites. Important things to think about are:

- Hazards inherent within the product;
- Hazards that may be introduced at the process step in question;
- Hazards that may increase at the process step in question.

The next step should be for the HACCP team to evaluate each of the dangers that have been found. By evaluating each one's probability of happening and impact severity, the goal is to determine which ones have the greatest influence on feed or food safety. While some practitioners think that using a basic model to score hazards is helpful, it is imperative to make sure that the most severe risks receive the most attention regardless of whether a risk scoring method is employed.

### *6.3.3.6. Determination of Control Measures*

Anywhere there is a hazard that poses a substantial danger, it is crucial to implement a control measure or steps to eliminate it or decrease it to a manageable level. The control measure(s) must be workable and feasible; however, they can take many different forms. Every control that is implemented needs to be verified to make sure it works. This could involve proving, for instance, by analysis or other methods, that a claim made regarding a control is accurate and that the control functions as intended. For future reference, records of this should be preserved.

### *6.3.3.7. Determination of Critical Control Points (Codex Principle 2)*

Critical Control Points (CCPs) are process steps that require control measures in order to prevent, eliminate, or decrease risks to an acceptable level (meaning the hazard would not be discovered or eliminated at any later point in the operation). These steps must be designated as CCPs in the HACCP plan. The HACCP plan flow diagram should prominently display the process step that has been designated as a CCP at its precise position inside the processing facility.

### *6.3.3.8. Establishing Critical Limits (Codex Principle 3)*

The critical limits for the control measures for each of the CCPs should be described in detail by the HACCP Team. The line separating what is acceptable from what is unacceptable is known as the crucial limit. Legislative mandates will define certain crucial boundaries, but scientific research or experience will determine others. The term "critical limits" refers to the highest and/or lowest value that a chemical, biological, or physical hazard must be controlled at a CCP to avoid, eradicate, or decrease its occurrence to a tolerable level.

### *6.3.3.9. Monitoring (Codex Principle 4)*

To make sure recognized dangers are kept under control, the CCPs in the operation as well as the feed products themselves should be inspected, sampled, and monitored. Monitoring systems ought to be built with the ability to detect inefficient controls as soon as feasible, before they fail. As a result, the frequency of any monitoring is crucial and must to be mentioned in the HACCP plan. The monitoring tasks outlined in the HACCP plan are carried out by appropriately qualified and authorized individuals. What will be monitored, how and where it will be monitored, how often it will be done, and who will do the monitoring should all be outlined in the HACCP plan.

### 6.3.3.10. Preventive/Corrective Actions (Codex Principle 5)

When information indicates that a control measure or measures are not operating within critical limitations, it is imperative that appropriate, timely, and effective remedial action be taken. Whatever steps are implemented ought to address the root of the issue as well as its effects. What should happen if a CCP spirals out of hand should be outlined by the HACCP team. A clear assignment of responsibilities and documentation is necessary for the implementation of remedial actions. It is imperative to guarantee that protocols take into account the necessary measures for each product that has been processed after controls were last verified to be functioning within reasonable bounds. Stock bonding or even a product recall from consumers or middlemen may be necessary for this. Every remedial activity needs to be recorded.

### 6.3.3.11. Verification (Codex Principle 6)

To make sure that everyone is following the HACCP Plan's standards and that the Plan is working (validation), the HACCP team needs to put verification measures in place. Verification systems need to check for compliance with the HACCP plan and that related records are reviewed. The HACCP Plan may include multiple control measures at a CCP, each with the proper monitoring. Principle 6 (verification) should comprise confirming that all monitoring tasks are carried out, operating within critical limits, and validating the effectiveness of the control measure. The following things need to be taken into account when creating verification systems:

- Sampling and testing;
- Complaints monitoring;
- Internal auditing of the HACCP system;
- External auditing of the HACCP system.

### *6.3.3.12. External Auditing of the HACCP System*

The HACCP team should conduct routine reviews to confirm that feed product safety is effectively and consistently ensured by the plan and that all requirements are being satisfied in practice. Every year, there should be at least one thorough HACCP review that takes into account any qualifications that were set down in the HACCP Plan. HACCP reviews should be documented, along with the conclusions reached by the HACCP team and any corrective measures taken.

### *6.3.3.13. Recordkeeping (COdEx Principle 7)*

Records serve as a method to detect possible issues as well as a way to track the product's history and documented proof that the HACCP plan is being followed. An overview of the hazard analysis, the HACCP plan, supporting documentation, and operating records are the four typical categories of HACCP records. The hazard analysis is summarized in the records. Documenting the HACCP team's deliberations is beneficial in supporting the team's conclusion regarding which risks are considered relevant to humans and are incorporated into the HACCP plan. Justification or description of the control measures that stop, remove, or lower the risk to a manageable level should be included in good documentation. The HACCP team, the feed and its distribution, the intended use, the consumers, and the process flow should all be listed in the HACCP plan as preliminary steps. Indicate on the process flow where the important control points are at each of the relevant process phases, and then include the HACCP summary table in your overall record keeping. Critical control points, monitoring, corrective action, verification, and critical limits are all included in the supporting documentation for HACCP concepts. SOPs that were created for control measures at the CCPs are among the other supporting documentation. In order to properly apply the HACCP plan, daily operational records must include monitoring, corrective action, and verification records. These documents demonstrate that the HACCP strategy is being implemented. The foundation of the HACCP system is monitoring records, which are intended to record plan compliance. Automation is probably used to record continuous monitoring. For instance, the mill automation system records the temperature of the conditioned mash. Standardized forms are necessary for reliable reporting of continuous monitoring. When monitoring detects a failure to meet a critical limit, corrective action documentation is initiated. Product codes and quantities that have been released, destroyed, or altered should be included in the records. Documentation is required for all verification operations specified in the HACCP plan, including calibration records, daily operation verification, any validation or reevaluation of the HACCP plan, and on-site audits.

## 6.4. ON-FARM PRODUCTION AND USE OF FEED

Feed safety is impacted by four aspects of farm operations: growing feed crops on-site (such as cereals, pulses, forage crops, and pasture); buying and using feed that is brought in; processing, mixing, and storing feed on the farm; and feeding livestock that will provide food for humans (such as meat, milk, eggs, etc.). Delivering the necessary nutrients to livestock at the lowest possible cost and preventing contamination that can compromise food safety is the aim of all of these endeavors. Profitability will increase along with animal performance when safe feed is produced and used. Getting safe ingredients is the first step because you cannot make safe feed without them. This section addresses the following topics: relevant components of appropriate animal feeding practices; feed manufacturing on-farm; and good agricultural practices for the production of feed.

### 6.4.1. GAPs

In its broadest sense, GAP refers to the application of existing knowledge to environmental, social, and economic sustainability issues for on-farm production as well as post-production procedures that yield food that is safe and healthful. A wide range of stakeholders' concerns and commitments regarding food production and security, food safety and quality, and the environmental sustainability of agriculture have led to a recent evolution of the concept of Good Agricultural Practices in the context of a fast-changing and globalizing food economy. GAPs are applicable to the main production line, which includes primary processing and crop material growth. The initial feed ingredient material is then submitted to manufacture and processing in compliance with GMPs. The dangers of chemical or microbiological contamination, improper use of crop protection chemicals, and deterioration during primary processing and storage are all eliminated or significantly decreased by GAPs. Reducing and eliminating these hazards improves the feed materials' dependability. Therefore, to lower the risks of contamination of the feed and food chain, feed materials and ingredients should be provided by suppliers who can adhere to Good Agricultural Practices. The following constitute Good Agricultural Practices:

1. **Agricultural Sites/Production Areas:** Once the production sites are appropriate and do not pose problems for crop contamination, operator health, or the environment, feed and food safety are safeguarded.
2. **Seeds and Propagation Materials:** Make use of high-quality seeds that are devoid of harmful pests, illnesses, viruses, and so on.

Growing seed varieties designed to be resistant to seed-infecting and insect pests is one recommended technique for the prevention and reduction of mycotoxin contamination in cereals, raw materials, and supplemental feedstuffs for milk-producing animals. In a given region of the nation, only seed kinds that are advised for use there should be sown.

3. **Crop Rotation and Soil Management:** Create and follow a crop rotation plan to prevent growing the same crop in the same field twice in a row. To lower inoculums in the field, crops that are not hosts to Fusarium species, such as potatoes, other vegetables, clover, and lucerne, should be utilized in rotation. It has been discovered that wheat and maize are especially vulnerable to Fusarium species, hence they shouldn't be utilized interchangeably.
4. **Use of Agricultural Chemicals:** Agricultural chemicals must be properly marked and purchased from reliable vendors. Agricultural chemicals should be kept out of the way of other goods and animal feed in clean, dry areas, in securely labeled containers. Agricultural chemicals such as herbicides, insecticides, fertilizers, and others should be applied for the intended purpose and in the amounts and intervals recommended by the manufacturer. All application justifications including the name and composition of the chemical utilized should be included in the records of the application.
5. **Use of Fertilizers:** Assess the need for fertilizers or soil conditioners to maintain a pH level that is appropriate for plants and to prevent plant stress, particularly during seed development. The application of either organic or inorganic fertilizers should only be advised by qualified individuals. Note the date, month, and year of fertilizer applications as well as the type and concentrations of the fertilizer.
6. **Manure:** When using manure for pasture and feed crops, it is important to handle and store it properly to reduce environmental pollution, especially from runoff that contaminates groundwater and streams. To reduce biological contamination and give the manure time to break down, there should be enough time between applying it and grazing. In a similar vein, manure added to ponds to increase production needs to decompose for a sufficient amount of time before being used to reduce the amount of germs present.
7. **Irrigation:** When using irrigation, be sure that all plants receive an equal amount of water. The quality of the water utilized should be sufficient. Chemical concentrations and other water quality parameters should be known when using wastewater; these can

be found in the WHO Guidelines for the Safe Use of Wastewater, Excreta, and Greywater (2006). The principles and processes for the safe use of wastewater, including minimum procedures and health-based goals, are outlined in these guidelines. It also addresses a significant update to methods for guaranteeing the microbiological safety of wastewater utilized in farming.

8. **Integrated Pest Management:** By combining current pest control techniques with environmental and pest information, integrated pest management, or IPM, aims to prevent unacceptable levels of pest damage while using the least amount of money and posing the fewest risks to people, property, and the environment.

9. **Plant Protection Products:** Plant protection solutions may need to be used when non-chemical methods and procedures are insufficient to manage pests. These should be suitable for the pests, diseases, and weeds under consideration, and they should be handled and kept correctly in accordance with label requirements. Use of plant protection products should only be restricted to those approved for use in the particular crop and in the nation in which they are registered.

10. **Harvest:** These crops when they are fully developed, as permitting them to reach maturity might expose them to severe heat, heavy rain, or drought.

11. **Storage and Distribution:** Crop residue from previous harvests should be removed from harvested areas for storage. To avoid insect infestation, storage facilities should, when suitable, be cleaned and treated with insecticide before being used. For commodities that are bagged, make sure the bags are clean and arranged on pallets, or place a layer of material that is impermeable to water between the sacks and the floor. Harvested crops should be maintained at the temperature that will prevent the growth of mold and insects while maintaining the product's physiological and physical integrity. Aerate bulk commodities whenever feasible to keep the right moisture and temperature. Employing appropriate and approved preservatives, like propionic acid or other organic acids, can be advantageous as they effectively eradicate mold and fungi and stop the formation of mycotoxins.

12. **Transport:** These containers ought to be dry and devoid of any apparent insects, fungi, or other contaminated materials. Transport containers must be appropriate for the intended cargo and cleaned and disinfected as needed before being used and reused. Using

pesticides or fumigants that are registered may be beneficial. The transport container needs to be completely empty of all cargo and cleaned as needed upon offloading. Grain shipments should be covered, sealed, or covered with tarpaulins to keep out extra moisture. Steer clear of temperature swings and other actions that can lead to condensation on the grain, which could result in an accumulation of localized moisture, fungal development, and the production of mycotoxin. Use insect and rodent-proof containers or chemical treatments that repel insects and rodents, if appropriate for the grain's intended end use, to prevent infestation by insects, birds, and rodents during transportation.

13. **Equipment:** All trucks, trailers, and other gear and equipment used in the transportation of grains and other feed raw materials should be cleaned. In order to avoid cross-contamination, exercise extra caution when cleaning trailers that are used to transport various commodities and medicated feed. Inform employees of all cleaning protocols and documentation requirements. Ingredients, pre-mixes, and bulk feed concentrates should not be loaded onto machinery that is also used to transport insecticides, pesticide-treated wood, glass, or scrap metal.

14. **Documentation and Record Keeping:** It is ensured that farmers have appropriately designed, implemented, and updated efficient feed production and management systems by documenting procedures and pertinent farm practices. It is possible to prove compliance with legal, regulatory, and client obligations by documenting the practices outlined in the procedures. Maintaining records will make it easier to trace goods and information, comply with regulatory requirements, conduct external audits and inspections, and make data available to the appropriate authorities.

15. **Personnel Health, Safety, and Training:** The effectiveness and security of agricultural output depend on the health, safety, and cleanliness of the workforce. Personnel that have received training and education will be competent to carry out their responsibilities and will be well aware of the conditions and hazards that could compromise the safety and quality of the products.

# Bibliography


1. Allison, J. B., & Anderson, J. A. (1945). The relation between absorbed nitrogen, nitrogen balance, and biological value of proteins in adult dogs: Two figures. *The Journal of Nutrition, 29*(6), 413–420.
2. Bender, A. E. (1961). Determination of the nutritive value of proteins by chemical analysis.
3. Bender, A. E., & Doell, B. H. (1957). Biological evaluation of proteins: A new aspect. *British Journal of Nutrition, 11*(2), 140–148.
4. Block, R. J., & Mitchell, H. H. (1946). The correlation of the amino-acid composition of proteins with their nutritive value.
5. Cherian, G. (2020). *A Guide to the Principles of Animal Nutrition*. Oregon State University.
6. Cole, A. S., & Eastoe, J. E. (1988). The metabolism of proteins and amino acids. In A. S. Cole & J. E. Eastoe (Eds.), *Biochemistry and Oral Biology* (2nd ed., pp. 273–288). Butterworth-Heinemann.
7. Derse, P. H. (1962). Evaluation of protein quality (biological method). *Journal of AOAC International, 45*, 418.
8. Dreyer, J. J. (1957). The nitrogen: Water ratios of albino rats and their use in protein-evaluation tests. *British Journal of Nutrition, 11*(1), 22–27.
9. Ezurike, B. O., Osazuwa, O. J., Okoronkwo, C. A., & Okoji, K. I. (2018). Design, construction and performance evaluation of a flat screen hammer mill machine. *African Journal of Science, Technology, Innovation and Development, 10*(6), 683–690.
10. Food and Agriculture Organization of the United Nations (FAO). (1957). *Protein Requirements*. FAO Nutritional Studies No. 16, Rome.

11. García-Yuste, S., & Pérez-Barbería, F. J. (2020). The ruminant: Life history and digestive physiology of a symbiotic animal. In *Sustainable and Environmentally Friendly Dairy Farms* (pp. 19–45).
12. Grzesiak, M., Knapczyk-Stwora, K., & Slomczynska, M. (2021). Vitamin D3 in ovarian antral follicles of mature gilts: *Expression of its receptors and metabolic enzymes, concentration in follicular fluid and effect on steroid secretion in vitro. Theriogenology*, 160, 151-160.
13. Grzesiak, M., Waszkiewicz, E., Wojtas, M., Kowalik, K., & Franczak, A. (2019). Expression of vitamin D receptor in the porcine uterus and effect of 1, 25 (OH) 2D3 on progesterone and estradiol-17β secretion by uterine tissues in vitro. *Theriogenology,* 125, 102-108.
14. Gupta, P. K. (2018). Epidemiology of animal poisonings in Asia. In *Veterinary Toxicology* (pp. 57–69). Academic Press.
15. Hegsted, D. M., & Neff, R. (1970). Efficiency of protein utilization in young rats at various levels of intake. *The Journal of Nutrition, 100*(10), 1173–1179.
16. Hong, S. H., Lee, J. E., Kim, H. S., Jung, Y. J., Hwang, D., Lee, J. H., ... & An, B. S. (2016). Effect of vitamin D3 on production of progesterone in porcine granulosa cells by regulation of steroidogenic enzymes. *Journal of biomedical research*, 30(3), 203.
17. Huang, H., Szumacher-Strabel, M., Patra, A. K., Ślusarczyk, S., Lechniak, D., Vazirigohar, M., & Cieślak, A. (2021). Chemical and phytochemical composition, *in vitro* ruminal fermentation, methane production, and nutrient degradability of fresh and ensiled Paulownia hybrid leaves. *Animal Feed Science and Technology, 279*, 115038.
18. Jang, H., Choi, Y., Yoo, I., Han, J., Hong, J. S., Kim, Y. Y., & Ka, H. (2017). Vitamin D-metabolic enzymes and related molecules: *Expression at the maternal-conceptus interface and the role of vitamin D in endometrial gene expression in pigs. PLoS One*, 12(10), e0187221.
19. Kim, H. J., Mittal, M., & Sonenshein, A. L. (2006). CcpC-dependent regulation of citB and lmo0847 in *Listeria monocytogenes. Journal of Bacteriology, 188*(1), 179–190.
20. Kundu, M. C., Wojtusik, J., & Johnson, P. A. (2012). Expression and regulation of Kit ligand in the ovary of the hen. *General and Comparative Endocrinology*, 179(1), 47-52.
21. Lalhriatpuii, M., & Patra, A. K. (2022). Feeding rice fermented beer waste improves growth performance, nutrient digestibility, and nitrogen utilization in growing rabbits. *Journal of Animal Physiology and Animal Nutrition, 106*(1), 147–155.

22. Lillehoj, H., Liu, Y., Calsamiglia, S., Fernandez-Miyakawa, M. E., Chi, F., Cravens, R. L., & Gay, C. G. (2018). Phytochemicals as antibiotic alternatives to promote growth and enhance host health. *Veterinary Research, 49*, 1–18.

23. Madeira, M. S., Cardoso, C., Lopes, P. A., Coelho, D., Afonso, C., Bandarra, N. M., & Prates, J. A. (2017). Microalgae as feed ingredients for livestock production and meat quality: A review. *Livestock Science, 205*, 111–121.

24. Miller, D. S. (1963). A procedure for determination of NPU using rats body N technique. In *Evaluation of Protein Quality* (pp. 34–36). National Academy of Sciences (NRC), Washington, DC, USA.

25. Miller, D. S., & Payne, P. R. (1961). Problems in the prediction of protein values of diets: The influence of protein concentration. *British Journal of Nutrition, 15*(1), 11–19.

26. Min, B. R., Parker, D., Brauer, D., Waldrip, H., Lockard, C., Hales, K., & Augyte, S. (2021). The role of seaweed as a potential dietary supplementation for enteric methane mitigation in ruminants: Challenges and opportunities. *Animal Nutrition, 7*(4), 1371–1387.

27. Mitchell, H. H. (1924). A method of determining the biological value of protein. *Journal of Biological Chemistry, 58*(3), 873–903.

28. Mitchell, H. H., & Carman, G. G. (1926). The biological value of the nitrogen of mixtures of patent white flour and animal foods. *Journal of Biological Chemistry, 68*, 183.

29. Mitchell, H. H., Beadles, J. R., & Kruger, J. H. (1927). The relation of the connective tissue content of meat to its protein value in nutrition. *Journal of Biological Chemistry, 73*, 767.

30. Mutwedu, V. B., Manyawu, G. J., Lukuyu, M. N., & Bacigale, S. (2020). *Fodder Production Manual for Extension Staff and Farmers in South Kivu and Tanganyika Provinces of the Democratic Republic of the Congo*. ILRI Manual.

31. Nasef, M., Sauber, A., Sakr, A., & Geba, K. (2020). An efficient model to encrypt text and gray image based on amino acid chains. *The Egyptian Journal of Language Engineering, 7*(2), 20–31.

32. Ogola, O. E. (2022). Physiological responses of broiler chickens to exogenous emulsifier supplementation in tallow-incorporated reduced-energy diets.

33. Osborne, T. B., Mendel, L. B., & Ferry, E. L. (1919). A method of expressing numerically the growth-promoting value of proteins. *Journal of Biological Chemistry, 37*(2), 223–229.

34. Ozturkoglu-Budak, S. (2017). A model for implementation of HACCP system for prevention and control of mycotoxins during the production of red dried chili pepper. *Food Science and Technology, 37, 24-29.*

35. Patra, A. K. (2010). Effects of supplementing low-quality roughages with tree foliages on digestibility, nitrogen utilization, and rumen characteristics in sheep: A meta-analysis. *Journal of Animal Physiology and Animal Nutrition, 94*(3), 338–353.

36. Patra, A. K., & Kar, I. (2021). Heat stress on microbiota composition, barrier integrity, and nutrient transport in gut, production performance, and its amelioration in farm animals. *Journal of Animal Science and Technology, 63*(2), 211.

37. Pérez-Barbería, F. J., Elston, D. A., Gordon, I. J., & Illius, A. W. (2004). The evolution of phylogenetic differences in the efficiency of digestion in ruminants. *Proceedings of the Royal Society of London. Series B: Biological Sciences, 271*(1543), 1081–1090.

38. Riet-Correa, F., & Medeiros, R. M. (2001). Intoxicações por plantas em ruminantes no Brasil e no Uruguai: Importância econômica, controle e riscos para a saúde pública. *Pesquisa Veterinária Brasileira, 21*, 38–42.

39. Robbins, C. (1993). *Wildlife Feeding and Nutrition*. Elsevier.

40. Rosentrater, K. A., & Evers, A. D. (2017). *Kent's Technology of Cereals: An Introduction for Students of Food Science and Agriculture*. Woodhead Publishing.

41. Roy, A., Mandal, G. P., & Patra, A. K. (2013). Evaluating the performance, carcass traits, and conjugated linoleic acid content in muscle and adipose tissues of Black Bengal goats fed soybean oil and sunflower oil. *Animal Feed Science and Technology, 185*(1, 2), 43–52.

42. Ruschkowski, S. R., & Hart, L. E. (1992). Ionic and endocrine characteristics of reproductive failure in calcium-deficient and vitamin D-deficient laying hens. *Poultry science, 71(10), 1722-1732.*

43. Sherasia, P. L., Garg, M. R., & Bhanderi, B. M. (2018). Pulses and their by-products as animal feed. *United Nations.*

44. Smolikova, K., Mlynarcikova, A., & Scsukova, S. (2013). effect of 1α, 25-dihydroxyvitamin D. *Endocrine Regulations, 47, 123-131.*

45. Stanton, T. L., & LeValley, S. B. (2006). *Feed Composition for Cattle and Sheep*. Colorado State University Extension Service.

46. Te Pas, M. F., Veldkamp, T., de Haas, Y., Bannink, A., & Ellen, E. D. (2021). Adaptation of livestock to new diets using feed components without competition with human edible protein sources: A review of the

possibilities and recommendations. *Animals, 11*(8), 2293.

47. Thomas, K. (1909). Über die biologische Wertigkeit der Stickstoffsubstanzen in verschiedenen Nahrungsmitteln.
48. Tomar, K., Tiwari, S., Singh, D., & Kumar, D. (2022). Feeding management of ruminant livestock. In *Smart Agricultural Technologies* (pp. 71–84).
49. Toral, P. G., Monahan, F. J., Hervás, G., Frutos, P., & Moloney, A. P. (2018). Modulating ruminal lipid metabolism to improve the fatty acid composition of meat and milk: Challenges and opportunities. *Animal, 12*(s2), s272–s281.
50. Urugo, M. M., Teka, T. A., Teshome, P. G., & Tringo, T. T. (2021). Palm oil processing and controversies over its health effect: Overview of positive and negative consequences. *Journal of Oleo Science, 70*(12), 1693–1706.
51. Van Eijkeren, J. C., Zeilmaker, M. J., Kan, C. A., Traag, W. A., & Hoogenboom, L. A. P. (2006). A toxicokinetic model for the carry-over of dioxins and PCBs from feed and soil to eggs. *Food Additives and Contaminants, 23*(5), 509–517.
52. Wojtusik, J., & Johnson, P. A. (2012). Vitamin D regulates anti-Mullerian hormone expression in granulosa cells of the hen. *Biology of reproduction, 86(3), 91-1.*
53. Yao, X., Wang, Z., El-Samahy, M. A., Ren, C., Liu, Z., Wang, F., & You, P. (2020). Roles of vitamin D and its receptor in the proliferation and apoptosis of luteinised granulosa cells in the goat. *Reproduction, Fertility and Development, 32(3), 335-348.*
54. Yao, X., Zhang, G., Guo, Y., Mohamed, E. S., Wang, S., Wan, Y., ... & Zhang, Y. (2017). Vitamin D receptor expression and potential role of vitamin D on cell proliferation and steroidogenesis in goat ovarian granulosa cells. *Theriogenology, 102, 162-173.*
55. Yucel, B., & Taskin, T. (2018). *Animal Husbandry and Nutrition.* BoD–Books on Demand.

# INDEX

Kolibri 360 GmbH
Ferdinand-Jühlke-Straße 7
99095 Erfurt, Deutschland
produktsicherheit@kolibri360.de